Software-defined Infrastructure

To my beloved mother, Aïcha Salah, who died in my arms on December 28th, 2024, leaving behind an immeasurable void. You have been an example to me, and always will be. You live on in each of us.
Abdelhamid Mellouk

To my two grandmothers, my beloved parents, my dear sister, and the one who quietly stood by my side throughout this journey: my deepest appreciation for your unconditional love, unwavering support, and endless encouragement.
Tran-Tuan Chu

To my family, my rock. To Alma, who changed everything.
Mohamed-Aymen Labiod

To my beloved family.
Brice Augustin

New Generation Networks Set

coordinated by
Abdelhamid Mellouk

Volume 4

Software-defined Infrastructure

A Novel Concept of Advanced Network Infrastructure

Tran-Tuan Chu
Mohamed-Aymen Labiod
Brice Augustin
Abdelhamid Mellouk

iSTE

WILEY

First published 2025 in Great Britain and the United States by ISTE Ltd and John Wiley & Sons, Inc.

ISTE Ltd
27-37 St George's Road
London SW19 4EU
UK

www.iste.co.uk

John Wiley & Sons, Inc.
111 River Street
Hoboken, NJ 07030
USA

www.wiley.com

Library of Congress Control Number: 2025935238

British Library Cataloguing-in-Publication Data
A CIP record for this book is available from the British Library
ISBN 978-1-83669-053-5

Contents

Preface

After two decades of development history, the number of Internet of Things (IoT) devices is growing dramatically. Along with the dramatic growth of the general IoT paradigm, a crucial branch has received significant research attention from the community, which is the healthcare Internet of Things (H-IoT). This particular branch results from the IoT's emergence in the eHealth domain, which allows for innovation in humanity's healthcare prospects.

In a broad field such as eHealth, H-IoT covers healthcare data transportation, data collection and storage. Notably, delivering vital healthcare data in real-time is an essential factor behind the success of applications built around the H-IoT paradigm, especially in emergencies. Hence, this particular characteristic demands the most attention from the communications infrastructure, with consideration toward an architecture specific to this type of application. In this respect, the evolution of the Internet operating model, impacted by new approaches such as network function virtualization (NFV), network slicing and the technological advances offered by 5G/6G networks, represents new challenges and complexities to be considered. Furthermore, such an evolution also introduces promising prospects in the network redeployment context of crisis or natural disaster situations.

Part of this book was initially based on the work done in the framework of Tran-Tuan Chu's PhD thesis, under the direction of Professor Abdelhamid Mellouk. It was then elaborated on to create a book that can provide a cutting-edge framework that can advance the use of the IoT, especially prioritizing supporting healthcare providers in urgent situations.

After studying the state of the art and addressing the current limitations in the literature, we designed a coherent framework based on an incremental approach, integrating three levels from the infrastructure to the terminal node. Our framework concerns the network infrastructure required to carry all traffic types in a mobile and constrained environment. At the top level, we consider the cutting-edge network infrastructure as a unified ecosystem – software-defined infrastructure – which can also deploy and extend its infrastructure. Thanks to this approach, we could adopt a unified optimization model built around each hardware block within this ecosystem and overcome the hardware heterogeneity imposed by the network equipment diversity. We explore our framework deeper by addressing communication infrastructure's dynamicity issues, particularly in migrating service for a mobile object, which could be a moving emergency vehicle, such as an ambulance or similar, to provide pre-hospital care for a patient. Last but not least, at the bottom level, we present a standalone solution embedded within the mobile object to optimize end-to-end latency. More precisely, we proposed a deep learning-based approach to select the optimal multipath scheduler that manages multiple network interfaces based on current network characteristics to choose the most suitable interface for sending the traffic. The proposed framework and the solutions it offers constitute essential components for our use cases. Thus, three enhancement levels facilitate the overall performance improvement of all functional blocks within the context of the H-IoT field, whose uncertainty and complexity characteristics are distinguished.

Beside introducing H-IoT literature landscape, this book also aims to inspire the research community toward a coherent development road map by sharing a common goal, developing software-defined infrastructure altogether.

It would have been impossible to give a complete bibliography and a historical account of the research that led to the present form of the subject. It is thus inevitable that some topics have been explored in less detail than others. The choices made reflect, in part, personal taste and expertise, and, in part, a preference for very promising research and recent developments in the fields covered by the book, highlighting the ongoing need for research and development in this vital area.

This book is a starting point, but also leaves many questions unanswered. We hope that it will inspire a new generation of investigators and investigations.

Finally, the authors hope that you will find this book interesting, while sparking novel and helpful ideas for your future work.

Tran-Tuan Chu
Mohamed-Aymen Labiod
Brice Augustin
Abdelhamid Mellouk
April 2025

List of Acronyms

5G	Fifth Generation Technology Standard.
AdaBoost	Adaptive Boosting.
AI	Artificial Intelligence.
ANN	Artificial Neural Network.
API	Application Programming Interface.
ASs	Autonomous Systems.
BER	Bit Error RATE.
CDF	Cumulative Distribution Function.
CNN	Convolutional Neural Network.
DDoS	Distributed Denial-of-Service.
DL	Deep Learning.
DRL	Deep Reinforcement Learning.
GADaM	Generic Adaptive Deep Learning-Based Multipath Scheduler Selector.
H-IoT	Healthcare Internet of Things.
HoL	Head of Line blocking.

HTTP	Hypertext Transfer Protocol.
HTTPS	Hypertext Transfer Protocol Secure.
ICMP	Internet Control Message Protocol.
IETF	Internet Engineering Task Force.
IoT	Internet of Things.
IPsec	Internet Protocol Security.
ISP	Internet Service Provider.
KNN	K-Nearest Neighbors.
LLDP	Link-Layer Discovery Protocol.
MAC	Message Authentication Code.
MEC	Multi-access Edge Computing.
ML	Machine Learning.
MLP	Multi-Layer Perceptron.
MNO	Mobile Network Operator.
MOS	Mean Opinion Score.
MPTCP	MultiPath Transmission Control Protocol.
NFV	Network Function Virtualization.
NOs	Network Operators.
OF	OpenFlow.
P2P	Peer-to-Peer.
PDR	Packet Delivery Ratio.

QoE	Quality of Experience.
QUIC	Quick UDP Internet Connections.
RF	Random Forest.
RTT	Round Trip Time.
SCM	Smart Connection Migration.
SDI	Software-defined Infrastructure.
SDN	Software-defined Networking.
SDWBAN	Software-defined Wireless Body Area Network.
SINR	Signal-to-Interference-plus-Noise Ratio.
SNR	Signal Noise Ratio.
SS	Scheduler Selector.
SSH	Secure Shell Protocol.
SSL	Secure Sockets Layer.
TCP	Transmission Control Protocol.
TLS	Transport Layer Security.
TLS/SSL	Transport Layer Security/ Secure Sockets Layer.
UDP	User Datagram Protocol.
VPN	Virtual Private Network.
WBAN	Wireless Body Area Network.

Introduction

Digital change has been particularly challenging in healthcare...

da Fonseca et al. (2021)

I.1. H-IoT context

The Internet of Things (IoT) is the technology that enables the link between the world's objects and their identities in the digital space. After two decades of development, 15 billion things will be connected to the Internet by 2025, according to CISCO. This indicates the IoT's development capability and exposes various potential issues and opportunities with such dramatic growth. The upward trend in the IoT's number of devices is predicted to continue increasing even further because of the technology's broad adoption in various industries: manufacturing, logistics, smart driving, smart homes, smart cities, etc.

Among numerous domains influenced by the IoT paradigm, there is a crucial domain that consistently improves the quality of human life and also receives significant research attention from the community, which is eHealth. The eHealth paradigm describes the healthcare services supported by digital processes, technology and communication. Throughout its history, the eHealth transition has significantly improved healthcare services – automating redundant tasks, unifying digital health records, increasing patient satisfaction, lowering costs, etc. – by digitizing healthcare procedures. However, various trivial tasks that need to be included require the next level of innovation, such as large-scale health monitoring, telemedicine and telehealth. The emergence of the IoT paradigm in the eHealth domain thus leads to a branch formulation in the literature: healthcare Internet of Things (H-IoT). The emergence of this domain promises the deployment of ubiquitous multilayer connected smart devices that enable potential solutions and open new opportunities for future use cases.

To investigate existing research challenges in the field, especially toward the eHealth use case, we conduct a thesis and present our work in this book.

I.2. Scope identification

In the H-IoT branch, the critical factor that fuels the success of the ecosystem is delivering vital healthcare data in real time, especially in urgent circumstances. The main reason is that the prehospital stage is vital to maintaining the patients' conditions before treating them at the healthcare facilities. During this phase, time and communication continuity are essential to allow on-the-scene healthcare providers to deliver proper first aid to patients. Therefore, we oriented our solutions first to cope with such critical circumstances to satisfy its strict communication requirements and preserve the holism of our proposed solutions. However, their applicable prospects are feasible and unlimited in the general IoT context or even in less competitive scenarios and other domains.

Generally, the system architecture in the H-IoT domain comprises three layers: Things, Communication and

Processing. Despite many existing studies, most have focused on the healthcare-related computing tasks at the Things and Processing layers. Some might investigate the Communication layer to improve the network performance. Such improvement might happen at the gateway endpoint but rarely influences the network ecosystem. The main reason for such limitation comes from a factor that could easily be ignored by researchers. Most of the research has always been conducted with a common presumption: the communication infrastructure is static throughout the communication process and could not be influenced by a third-party actor.

However, this assumption might no longer fully reflect the typical network conditions. The principal factors leading to this situation are the complex Internet operations, the software-defined network's advancements, the rise in mobile users and the global deployment of the 5G network. To be more explicit, the complex Internet operations and the rise in mobile users cause variations in the network infrastructure while the transmission is happening. On the other hand, the global 5G deployment and SDN developments enable advanced features such as network function virtualization and network slicing. These features allow third-party actors to create a separate network from the Internet infrastructure. They could also join, operate, manage and share their assigned infrastructure with other participants. In short, the prospect of the Internet and the features it brought with it have evolved through time and invalidated the presumed legacy of a static H-IoT communication process. Acknowledging the situation, we are motivated to update the H-IoT's Communication layer with the latest network advancements (multiple network interfaces and cutting-edge transport protocol – QUIC). The following points specify our objectives in this work:

– We confirm the statements above by studying the H-IoT state of the art (SOTA).

– We present to the community our studied perspective, point out the existing limitations in the literature and later navigate our research directions.

– We also present our proposed solutions to deal with the existing drawbacks and fill the pointed-out research gap in the domain's most challenging conditions.

I.3. Main contributions

As mentioned in the previous discussion, our objective is to renovate the Communication layer in the H-IoT domain. Therefore, this book studies the latest opportunities offered by cutting-edge network technology. Briefly introducing our novel solutions, the following top-down view summarizes our contributions in this book:

– Software-defined infrastructure (SDI): a novel paradigm attempts to deal with network hardware heterogeneity by unifying the network infrastructure into one coherent logical view. The paradigm can provide dedicated solutions for critical network use cases, especially emergency H-IoT, by its central management capability and universal accessibility to the network devices. We formulated a unified functional model that allows the proposed paradigm to optimize the overall network performance or prioritize critical traffic in urgent situations. We also present four experimental approaches to evaluate the proposed paradigm's efficiency. Our obtained simulation results also confirmed the paradigm's network orchestrating capabilities and its effectiveness in improving network performances for vital data transmission.

– SMART connection migrations (SCM): an approach that takes advantage of the unified logical network overview empowered by the SDI paradigm to maintain a seamless network experience for emergency vehicles to transmit vital healthcare data. The proposed method not only enables the ability to ensure vital transmission continuity and optimize overall latency for onboard devices to edge computing servers, but also succeeds in reducing the required duration to execute the mobile device's connection handover in a simulation environment by 50%.

– Generic adaptive deep-learning-based multipath scheduler selector (GADaM): an intelligent approach that manages multiple network interfaces to provide the best network performance for the network terminal in highly dynamic network conditions without the SDI controller's assistance. In the simulation and practical environments, the "scheduler selector" deep-learning model achieved a high performance in selecting the most suitable scheduler based on multiple interfaces' network characteristics. In particular, the practical results also proved the model's capability to outperform the SOTA in some specific scenarios without prior knowledge.

I.4. Content organization

Having discussed the contributions of the book, we now present its organization. First of all, Chapter 1 covers the H-IoT SOTA. This chapter analyzes and classifies the surveyed papers to present the lessons learned and the opportunities neglected in the literature. Afterward, considering the given context and assisting the reader's navigation, we structure our contributions into three dedicated chapters regarding a general system's top-down view.

After the first chapter's overview, Chapter 2 introduces the first contribution of SDI, which influences the whole healthcare network infrastructure with its top-down overview. Initially, we present the literature's limited vision of the software-defined paradigm. Then, we investigate the research gap in Chapter 1. Later, we construct a new paradigm to fill the missing gap and enable future research. We also present a unified functional model with specified converged conditions and a simulation framework to concretize our proposed paradigm for prospective performance evaluations. Four distinguished experimental scenarios dedicated to supporting healthcare transport are presented to evaluate the novel paradigm's performance. The results confirm SDI's effectiveness by significantly improving the selected network metrics (bandwidth, delay and packet loss) versus the SOTA in healthcare vital data transmission scenarios.

Taking advantage of SDI's accessibility and knowledge about the healthcare network ecosystem, Chapter 3 presents an intelligent approach toward seamless and transparent network experiences for a healthcare endpoint: SCM. The approach aims to bridge the gap in the SOTA Internet connection concept that primarily focuses on static infrastructure but not highly dynamic and flexible network infrastructure, whose criteria are mobile healthcare transport scenarios. By employing SDI's network overview, our proposed approach promises to harness the cutting-edge Internet connection concept to maintain a seamless network experience and the feasibility of migrating service to optimize overall latency to the edge computing server for the terminal node. We also provide initial evaluation results in this chapter to confirm the improvements of the new approach versus the SOTA approach.

Modern network devices are equipped with multiple network interfaces thanks to cutting-edge technology development. However, these advantages have yet to exploit the offered opportunities fully and are adaptable in highly dynamic and complex environments, especially in healthcare transport scenarios. Hence, we present a novel concept that deals with such dynamicity by suggesting the most suitable scheduler based on the device's current network characteristics. Chapter 4 discusses the proposed *scheduler selector* paradigm and its proof-of-concept instance (GADaM) toward an independent solution for end-user devices in high dynamic environments. This chapter presents the arguments behind the paradigm and various simulation (accuracy metrics) and practical experimental (network metrics) results to validate our idea's efficiency. The results proved the proposed paradigm's capability to enhance the multi-network interface solutions in the mobile network environment.

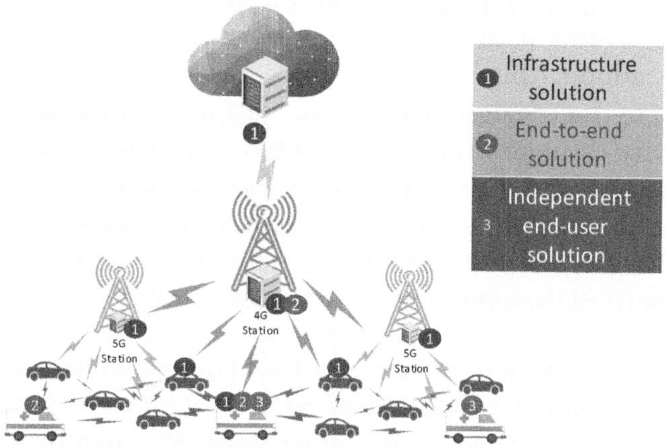

Figure I.1. *Contribution roadmap. For a color version of this figure, see www.iste.co.uk/chu/softwaredefined.zip*

For a better illustration, we also categorize our solutions by their influenced positions in a healthcare emergency framework. Figure I.1 describes the critical healthcare communication situation where an ambulance carries its patient to the hospital while transmitting vital information to the doctor and receiving the best instructions. Consider the first contribution: since it is essential to implement an SDI solution in each infrastructure's network element to enable its maximum capability, we classify it as an infrastructure solution. Next, the SCM feasibility only requires the implementation of the two specified endpoints. Hence, we refer to it as an end-to-end solution. Finally, because of GADaM's independent operation, it is only necessary to implement the third solution in the specified ambulance or end-user devices. Thus, we consider it an independent end-user solution. From the bottom-up angle in the ambulance's point of view, the impact of our proposed solutions to the H-IoT Communication layer could be described as follows. First, the ambulance uses GADaM to take advantage of multiple network interfaces to perform best in its critical situation. During its movement to carry the patient to the hospital, the ambulance adopts the SCM to ensure a seamless network experience to its designated server, despite unpredictable infrastructure changes (ambulance mobility, hardware failure, etc.). To enable the full potential of connection migration for the ambulance and assist its communications in the most challenging conditions, SDI's network overview and accessibility to SDI-enabled devices are employed.

Finally, the Conclusion and Prospects presents the final description for the combination of each contribution toward a coherent solution for an H-IoT emergency. We also present our prospects for the future work regarding each contribution and the projected advancements in this chapter.

Chapter 1

Healthcare Internet of Things: The State of the Art

The Internet of Things has the potential to change the world, just as the Internet did.

Kevin Ashton

This chapter presents an overview of the latest technological advancements in the healthcare Internet of Things (H-IoT) domain.

This work studies the opportunities offered by the most cutting-edge technologies to deal with challenges of H-IoT in the Communication layer. The following contents cover H-IoT's state of the art. First, section 1.1 presents the general H-IoT network's landscape in both software and hardware aspects. Second, section 1.2 describes our focused technology's emergence in the SOTA architecture in detail. Next, section 1.3 demonstrates our learned lessons and points out neglected opportunities offered by the network advancements for H-IoT, which later leads to our proposed solutions. We also position our research, define our concentration in this work, and present the prospects of our proposed solutions in this discussion. Finally, concluding this chapter, section 1.4 summarizes our findings and navigated directions.

1.1. H-IoT network landscape

In 1999, "Internet of Things" was first introduced by Kevin Ashton. The terminology describes a system in which objects in the physical world could be connected to the Internet by sensors. After more than 20 years of development history, IoT has become one of the crucial business market (Rejeb et al. 2022).

Along with the paradigm's advancements, a vital branch, H-IoT, has also progressed in the general IoT landscape, while sufficiently improving human life quality. Since it is a subset of the IoT paradigm, H-IoT architecture inherits the standard components in a general IoT architecture (Bouazizi et al. 2017; Elayan et al. 2017). Despite sharing common characteristics, Healthcare IoT also possesses some distinguish requirements compared to the generic IoT that specified by Qadri et al. (2020) in Table 1.1.

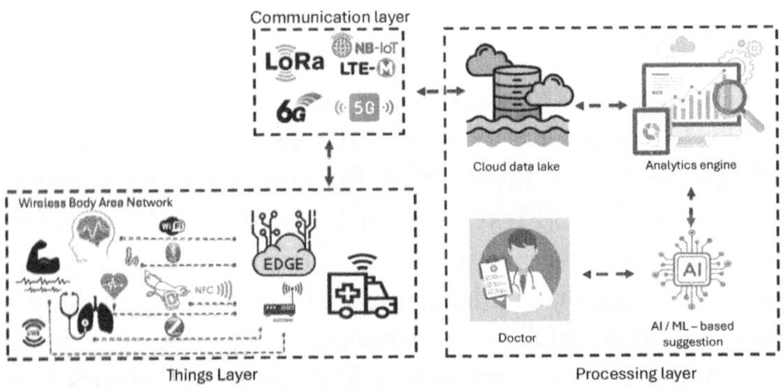

Figure 1.1. *Three-tier H-IoT architecture: Things, Communication and Processing layers. For a color version of this figure, see www.iste.co.uk/chu/softwaredefined.zip*

	Generic IoT	H-IoT
1	It is usually deployed over a large geographical area and serves a single purpose.	Usually deployed in a closed or small geographical area, either in and around the human body or in a healthcare facility.
2	Energy sources can include solar and wind energy. In case of stationary nodes, the nodes may be powered continuously.	H-IoT nodes can also harvest energy from human body using heat, stress and motion.
3	Monitor environment, used in defence applications, industrial monitoring.	Used to monitor human body vitals.
4	Desirably small nodes but node size vary according to environment and application.	The nodes are miniaturized to be unobtrusive.
5	Usually these are stationary.	Essentially mobile as associated with human body.
6	Deployment of sensors is comparatively easy.	Deployment is a difficult particularly in case of implants which mostly requires a surgery.
7	Data integrity is tried to maintain. The redundancy compensates for errors.	The data must be preserved and transmitted with utmost integrity.

Table 1.1. *Healthcare IoT and generic IoT comparision*

Besides identifying H-IoT's characteristics from general IoT, Qadri et al. (2020) also presented an abstract H-IoT communication architecture.

Inspired by their work, we update the general H-IoT architecture with the most recent advancements, such as the adoptions of 5G/6G, Cloud Data Lake and AI intervention in supporting Healthcare decision making, and illustrate these changes in Figure 1.1. The architecture still indicates the fundamental healthcare data flow as it consisted of three main layers: the Things, Communication and Processing layers.

Technology	Types	Frequency	Data Rate	Range	Power Usage
NFC	PAN	13.56 MHz	100–400 kbps	10 cm	Very Low
Bluetooth 4	PAN	2.4 GHz	1 Mbps	0.1 km	Low
Bluetooth 5	PAN	2.4 GHz	2 Mbps	0.25 km	Very Low
ISO/IEC 15693	PAN	3.56 MHz	6.6–26 Kbit/s	1–1.5 m	Very Low
Z Wave	LAN	968–908 MHz	100 kbps	100 m	Very Low
RFID	LAN	13.56 MHz–2.45 GHz	40–640 kbps	1–100 m	Low
Thread	LAN	2.4 GHz	250 Kbits/s	10–100 m	Low
Wi-Fi	LAN	2.4 GHz and 5 GHz	802.11(b)11 M; (g) 54 M; (n) 0.6; (Gac) 1 Gbps	50 m	Low–High
ZigBee	LAN	2.4 GHz	250 kbps	10–100 m	Very Low

Table 1.2. *Short range communication technologies parameters*

The Things layer collects the vital metrics generated by the sensing devices and then forwards those toward higher processing units (gateway, edge/fog node) for further processing. Since the sensing devices in the Thing layers operate close to the human body, any health-affected impact requires careful and standardized consideration. The devices' size, flexibility, and thermal output are crucially important for their implementation feasibility. With limited power, memory and computing capacity, the sensing devices required a gateway to contain the data and later transmit the collected data to the central server. Wired and wireless solutions have been developed over the years to transfer the collected data from the sensing devices to the gateway in the H-IoT branch. Initially, wired solutions have been developed and utilized on the human body in healthcare facilities and the laboratory environment. However, these solutions have disadvantages (Talpur et al. 2019) such as the amount of sensors' limitation and inconvenience placement on the body and causing barriers in data collection. Hence, they are only suitable for a few specific use cases.

On the other hand, wireless solutions also known as wireless body area networks (WBAN) resolved most aforementioned disadvantages. Besides, the technology's advancements also help to reduce the size of the sensing device's communication component. Thus, the sensing device could be easily integrated within the human body while allowing various vital metrics to be transmitted to the gateway over the air (Talpur et al. 2019). Furthermore, since the operational environment of WBAN is the human body, most WBAN communication technologies fall in the short-range communication category. Well-known WBAN technologies and their communication parameters are in Table 1.2 (Ahad et al. 2020). These characteristics are also the leading factors toward the dominance of WBAN technology in the Things layer in the H-IoT branch. After collecting the data successfully, the gateway could preprocess it to reduce the computing and network resources later or forward it directly to the Processing layer via the Communication layer. Such edge interventions provide real-time critical analysis and/or the fastest response, which are critical for the H-IoT use case.

After the preprocessing phase, the data continue its lifecycle and is transmitted to the Processing layer via the communication infrastructure. The gateway indeed requires a higher computing and power capacity and long-range communication technologies to transfer the data to the Processing layer.

Long-range communication technologies suitable for this purpose are presented in Table 1.3 (Ahad et al. 2020). Once the data pass through the Internet infrastructure and are handled by the central data center, the final processing at the Processing layer occurs. At the end of the data lifecycle, the computed results or any necessary instructions would be returned to the Things layer in the reverse direction, and a new data lifecycle begins.

Technology	Types	Frequency	Data Rate	Range	Power Usage
WiMAX	WAN	10–66 GHz	11–100 Mbs	50 km	High
LoRa	WAN	868/915 MHz	50 kbps	25 km	Low
LoRaWAN	WAN	Numerous	0.3–50 kbps	2 – 5 km (Urban) 15 km (Suburban) 45 km (rural)	Low
Sigfox	WAN	868/915 MHz	300 bps	50 km	Low
4G	WAN	700, 1700, 2800 MHz	Up to 12 Mbps	Up to 10 km	High
5 G	WAN	At Low Bands	Up to 3.6 Gbps	Up to 10 km	High
5G	WAN	At High Bands	10 Gbps	< 1 km	High
(NB-IoT)	WAN	850 MHz	245 kbps	Up to 35 km	High
(EC-GSM IoT)	WAN	890 MHz	Up to 140 kbps	Up to 100 km	High
LTE-M (M1)	WAN	700, 1450–2200, 5400 MHz	0.144 Mbps	35 km	High

Table 1.3. *Long-range communication technologies parameters*

A highlight that should be noticed in Table 1.3 is 5G technology. Mentioning the decade science breakthrough, 5G is one of the best candidates for the spotlight. After years of being developed and standardized since its first practical trials (Morgado et al. 2018), 5G has been widely deployed all over the world. Indeed, the 5G's first highlight is enhanced mobile broadband (eMBB). Thanks to its incredible theoretical 10–30 Gbps data rate (Ahmad et al. 2023), a single 5G link speed overfulfills most challenging network requirements, including ultra high-definition video, video calling (Sudhamani et al. 2023) and extreme AR/VR (2.35 Gbps) (Mangiante et al. 2017). Hence, they allow for the feasibility of such applications in optimal operating environments. Massive machine type communication (mMTC) is 5G's second prospective advantage compared to 4G. The second advancement enables billions of smart devices to connect to the Internet (Attaran 2023). Along with eMBB, massive multi-input multi-output technology is exploited to increase the device deployment maximum density to 1 million devices/km^2 (Eswaran and Honnavalli 2023). Finally, ultra reliable and low-latency communications (uRLLC) is another crucial 5G signature feature. Its promised advancement is the minimal end-to-end latency, only 4 ms compared to 200 ms in 4G.

With such advantages, 5G has strengthened the IoT's wide adoption (Liu et al. 2018b) in various domains, and one of them is smart healthcare (Li 2019) or eHealth. In other words, 5G is a vital factor in recent H-IoT revolution (Ahad et al. 2020). As compliance in the state-of-the-art H-IoT architecture and worldwide 5G adoption, Ahad et al. (2020) presents one of the latest general H-IoT communication infrastructure architectures. Inspired by their work, we illustrate the cutting-edge H-IoT architecture in Figure 1.2 to highlight the device-to-device links, 5G's impact and emphasized the ambulance's role. Instead of describing the data flow like in Figure 1.1, this diagram indicates H-IoT hardware infrastructure and the interactions within the communication ecosystem. In addition to the Cloud computing infrastructure and medical care center being specified as the final destination in the Processing layer, various cellular technologies empower the communicating function for the Communication layer. Conversely, the people, ambulances, and Edge clouds act as the gateway endpoints in the Things layer. Besides, the device-to-device links also play a crucial role in inter-WBAN communication. Last but not least, the figure also presents a perspective of multi-hop metric transmission in the H-IoT domain.

Besides being a performance boosting factor to the Communication layer, 5G technology also unlocks a novel network management perspective to the public, which is network slicing (Barakabitze et al. 2020). Specifically, before 5G's existence, the community commonly delegates the infrastructure network management to the mobile network operators, Internet service providers, etc. There are two main reasons for this observation: the exclusive management of the industrial companies and the costly deployment and maintenance expenses of self-managed private networks. Network slicing in 5G, however, reshaped this network phenomenon. The core concept behind the paradigm will be discussed in detail in the next chapter. However, a point worth mentioning is the paradigm's inspiring impact. The slice concept offers the third-party actors to reflect their demands over their assigned overlay of a deployed network infrastructure. Therefore, a third party could operate their dedicated network overlay while the underlying network infrastructure are still maintained by the infrastructure providers. Thus, both aforementioned barriers are eliminated. Hence, 5G established a network perspective that enabled a plethora of opportunities for future research in the eHealth domain with an unconstrained hardware management landscape.

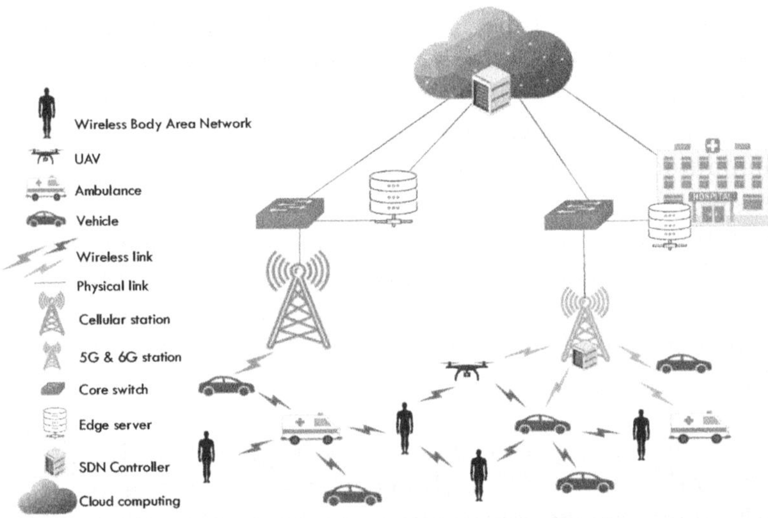

Figure 1.2. *H-IoT communication infrastructure. For a color version of this figure, see www.iste.co.uk/chu/softwaredefined.zip*

However, such promising prospects still require further effort to be achievable. One of the vital factors that held the community back in this direction is the limited number of network slices in the current 5G implementation. Despite numerous promising possibilities, only three slices are ready to use in 5G's implementation: eMBB, uRLLC and mMTC

(Popovski et al. 2018). Regarding their names, these slices' functions defined by the 3GPP standard are as follows. While eMBB slice aims to support a stable channel with high peak data rate, mMTC slice focuses on ensuring a massive number of IoT connections with small data payload. Last but not least, uRLLC allows 5G to achieve an ultra-low latency transmission (4 ms). Notably, each slice's objective is to satisfy a distinguished technical aspect defined by the infrastructure orchestrator. Hence, the specialized prospects for more complex and critical situations that require a mixture of the mentioned objectives still need to be clarified.

Correlated with this work, consider the crisis contexts or disaster scenarios as an example: eHealth communication is crucial to ensure the operation's success. However, since eHealth communication requires the most demanding communication specifications, it is vital to prioritize the medical task force's communication while maintaining average communication in such circumstances. Thus, constructing a temporary dedicated slice with a higher priority that satisfies eHealth communication constraints could be a feasible situational approach. Figure 1.3 is inspired by a use case of a 5G e-health slice in Thales' vision[1]. The figure illustrated the network-slicing vision of a customer's complex mixture requirements in a static industrial network infrastructure. First, it categorized three main layers in the network slicing concepts: tenant point-of-view; service orchestrators; and network infrastructure. Toward a unified hardware platform, the mobile transport and the hardware infrastructure perform the fronthaul and backhaul network integration between various infrastructure providers. On top of the common infrastructure, the slice orchestrator manages the united networking capability and computing resources to formulate network slices regarding

1 Presented by Emmanuel Dotaro, Head of ICT & Security labs, Thales/SIX/CTO, "5G basics and a bit beyond".

the customer's service-oriented specifications (mobile (virtual) network operator (MVNO), eHealth, automotive, media and entertainment). The slice orchestrator also provides a frontend, which serves as an entry point to support their tenant a service point-of-view for slicing creation and management requests. Even if the described prospect is practical, the current network slicing landscape is still tied to a predeployed network infrastructure. On the other hand, particularly in the eHealth domain, the infrastructure's integrity is uncertain in the mentioned challenging circumstances. A part or all of the existing infrastructure may malfunction or become unavailable because of the objective environmental causes (Khan et al. 2022) in such scenarios. Thus, a more versatile networking solution is expected to provide the communication capability for such time-critical situations. Once again, thanks to wireless communication advancements, flexible network coverage deployment is more feasible in the cutting-edge 5G network context. However, further investigation in this direction will be discussed in the next chapter. We have now provided the reader with the latest H-IoT network landscape. In the following discussion, we study the emergence of cutting-edge technologies in the H-IoT branch.

1.2. Technology emergence in H-IoT

After architectural-level and cutting-edge network landscape investigations, further discussions are emerging concerning the latest technologies in the H-IoT domain. The section aims to provide the audience with the overall landscape followed by this thesis navigation process before proceeding with our proposed contributions in the following chapters. To achieve this goal, we present two distinguished studies about our focused technology emergence in the H-IoT domain: multi-access edge computing and software-defined networks.

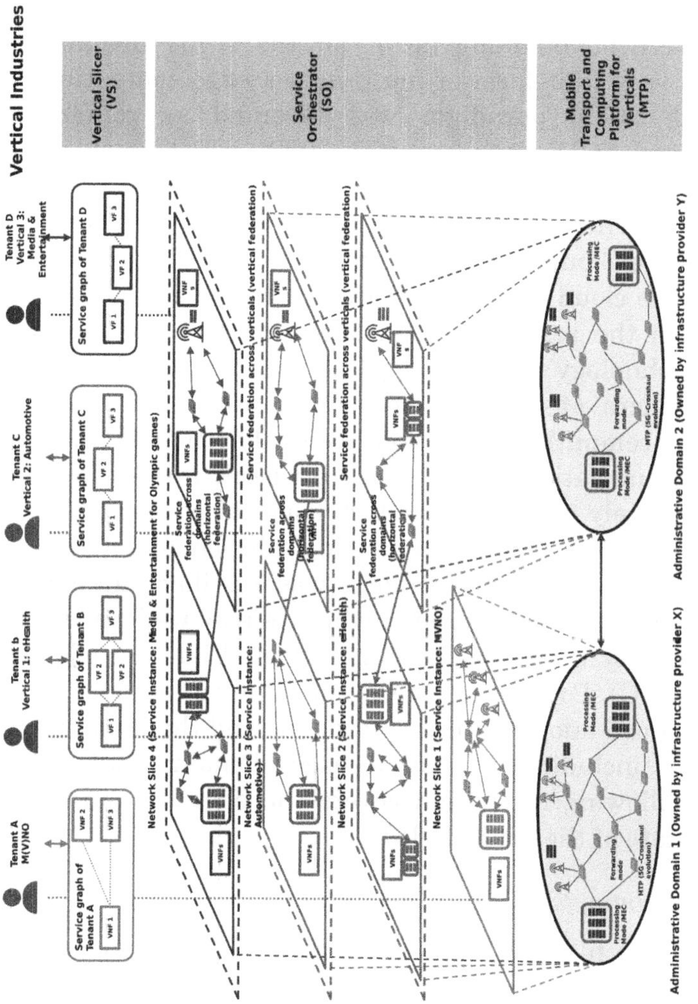

Figure 1.3. *eHealth dedicated slice technological landscape. For a color version of this figure, see www.iste.co.uk/chu/softwaredefined.zip*

1.2.1. *Edge computing in H-IoT*

1.2.1.1. *Cloud/fog/edge history*

Along with the described WBAN progression in the previous section, the cloud/fog/edge computing developments are another crucial enabling factor for the H-IoT domain. In 1990, the very first form of multi-access edge computing (Khan et al. 2019) paradigm was invented as content delivery network (CDN) (Dilley et al. 2002), which is about reducing the proximity between the end-users device and the web server. Then, in 2006, once Amazon promoted their Elastic Compute Cloud Service for the first time (Al-Dhuraibi et al. 2018), the cloud computing paradigm received massive attention from the researchers. In the middle of the cloud computing technology's embracing, Cisco introduced the fog computing (Hazra et al. 2023) paradigm in 2012 to cope with the cloud computing limitations (central bottleneck, long physical communication distance, etc.). Cisco's introduction not only effectively presented its technological edges to the community but also promoted an existing solution advantage, which is CDN. Later, the paradigm was justified, unified and named after its characteristic, which is now well known in the community under the terminology multi-access edge computing (MEC). Possessing a few crucial advantages, edge computing's broad adoption in the academic and the industry is inevitable, including in our focused domain, that is, H-IoT. The following discussion covers the edge computing paradigm's state of the art in H-IoT.

1.2.1.2. *Edge computing's taxonomy in H-IoT*

The first step to novel technology adoption in a domain is studying the domain's technological landscape at the architectural level. Various investigations in this direction have been published over the years.

In 2015, Ramalho et al. (2015) presented their progressive work toward the Multi-eHealth Cloud Service Framework

(MeCa). The novelty in this continuous work is migrating MeCa computing scenarios from the cloud environment to the fog/edge environment, highlighting the edge computing performance improvement.

To investigate this crucial advantage further, Vora et al. (2017) studied edge computing advantage in neurological treatment where real-time monitoring is essential. The authors presented an architecture to monitor patients with chronic neurological diseases and performed data clustering to detect anomalies before transmitting data to the destination.

Later, in 2018, Rais et al. (2018) proposed a four-tier architecture for a Healthcare monitoring system by integrating WBAN sensors, fog/edge computing, and cloud services, especially in an IPv6 environment. A small-scale testbed results confirmed the medical communication requirements satisfied by their proposed architecture.

Instead of the small-scale laboratory environment, Muhammed et al. (2018b) also presented a ubiquitous edge computing-based healthcare framework for smart cities. Various cutting-edge technologies are also adopted in this work, such as deep learning and big data, to provide a personalized healthcare networked experience. However, the infrastructure for such advancements could only be formulated with the edge computing paradigm. The results obtained from their proof-of-concept system confirms the efficiency in latency reduction (50% compared to traditional cloud-based healthcare system). The future work specified by the authors is improving the formulated framework's security, privacy, reliability and scalability.

Exploiting further the strategic position in the edge network of the intelligent gateway, Rahmani et al. (2018) investigated a range of high-level medical data processing services. Later on, their proof-of-concept implementation was adopted in several gateways to form an intermediary processing layer for an early warning healthcare monitoring system.

On the other hand, Greco et al. (2019) focused on medical data streaming tasks in the H-IoT context. The four-layer architecture's performance is also conducted by performing anomaly detection tasks in the REALDISP dataset with the latest technological framework. The obtained result confirmed the performance satisfaction of the designed system.

Meanwhile, Chen et al. (2018) explored the edge-based resource allocation strategy of the edge computing paradigm. The authors proposed an edge cognitive computing-based healthcare system. Taking advantage of computing resource allocations, the results, based on the data-driven cognitive approach at the edge, confirm the overall QoS improvement.

In addition to SOTA, Pace et al. (2018) developed a valid and inexpensive gateway solution for H-IoT. The two most significant advantages were the solution's capability to support multi-radio, multi-WBAN communication technologies and exploit both public and private Cloud platforms to guarantee the system's flexibility. Two case studies (factory workers and athletes training) were also defined for the evaluation phase.

Foreseeing LPWAN advantage in H-IoT deployment, Queralta et al. (2019) and Kharel et al. (2019) proposed an LPWAN edge computing based healthcare monitoring architecture. However, these two papers concentrated on different aspects. In Queralta et al. (2019), the authors focused on achieving high accuracy in a fall detection task and waiving the need to transmit raw data to cloud servers for online analysis in the proposed edge-intelligence integrated system. Meanwhile, Kharel et al. (2019) investigated physical connectivity metrics to ensure the feasibility of LPWAN-based H-IoT systems.

Toward combining (Pace et al. 2018) short-range with Queralta et al. (2019) and Kharel et al. (2019) long-range communication capacity, Wu et al. (2019) presented a hybrid H-IoT network system for industrial safety monitoring applications. Hence, a robust H-IoT hybrid solution, which employed the edge computing heterogeneous communication technology to reduce the latency and improve network performance, was conducted.

Considering the vast amount of data generated by the H-IoT monitoring system, Abdellatif et al. (2020) proposed an edge-computing architecture that provides efficient data delivery, namely, multimodal data compression and edge-based feature extraction for event detection. The studies investigated data distortion ratio, data classification accuracy and energy consumption to ensure the proposed architecture's short response time, efficient processing and minimal energy and bandwidth consumption. Not submitting a novel architecture, however, Satija et al. (2017) and Siddique et al. (2018) also showed their research interest in data filtering at the edge level. While Satija et al. (2017) viewed the problem from a classification point of view, Siddique et al. (2018) approached the problem via a data compression landscape. These studies confirmed the system's energy efficiency in transmitting acceptable quality data. Besides being concerned about qualified data transmission, Hayajneh et al. (2019) fulfilled the data protocol research gap. Their study was composed of three distinct algorithms toward three different protocol objectives. The channel assignment algorithm is for minimizing and avoiding interference. The gateway association algorithm aims to improve the connection's efficiency and security. Last but not least, their proposed routing metric is for constructing minimized delay and conserved power routes.

Along with communication architectural and general data processing contributions, a plethora of papers were conducted to study edge-computing IoT systems focusing on some specific health-based purposes. Specifically, they are diabetes detection (Priyadarshini et al. 2018; Devarajan et al. 2019; Abdel-Basset et al. 2020), chikungunya outbreak detection (Sood and Mahajan 2018), activity classification (Pham et al. 2018; Uddin 2019; Ram et al. 2019), Covid detection (Hegde et al. 2020), arrhythmia anomaly detection (Azimi et al. 2019), voice disorder (Muhammad et al. 2018a), skin region classification (Dai et al. 2019), stress-type classification (Priyadarshini et al. 2018) and food classification for diet assessment (Liu et al. 2018a). Since this chapter focuses on the complex H-IoT management aspect, we will not investigate this health-purpose specific direction in detail.

After the brief introduction to the state of the art, we refined the conducted knowledge and summarized it into our taxonomy in Figure 1.4 and Table 1.4.

1.2.2. *Software-defined network in H-IoT*

1.2.2.1. *Software-defined network history*

After investigating edge computing's emergence, we shift our focus to the next worth-mentioning emergence in H-IoT, that is, software-defined networks. If the edge computing paradigm represents the processing transition of Internet infrastructure (from cloud to fog then edge), the software-defined network is the representative of the Internet's management transition.

To have a clearer understanding of this particular transition, Vaniukov and Vaniukov (2023) may provide valuable insight into Internet's current operation. The article reports the enormous amount of money spent on IT infrastructure, which indicates how costly and complex to maintain and operate the Internet.

Reference	Contribution category	Data type or data set	Main Objective of exploiting edge computing in healthcare IoT	Performance metric	ML-based approach	Use case
Ramalho et al. (2015)	Framework	eHealth notification	Save energy, lower latency, avoid centralized penalties by Fog distributed computing	Received transmitted traffic, Energy consumption	NA	Alarm system for healthcare monitoring
Vora et al. (2017)	Framework, Implementation	Movement traces	Provide minimum delay to the queries	Response time, observed traffic, jitter, Energy consumption	K-mean , C4.5 decision tree	Patient monitoring system for ambient assisted living
Satija et al. (2017)	Signal quality classification	MIT-BIH arrhythmia, Physionet Physiological	Filter data and improve data quality, battery time of IoT devices, reduce the cloud server traffic load, bandwidth and treatment costs	Classification accuracy, computational time	NA	Continuous cardiac health monitoring
Rais et al. (2018)	Architecture, Implementation	Physiological	Reduce load and time bound	Delay	NA	Patient health monitoring in hospital environment
Jangra and Gupta (2018)	Framework	Time-series data	Lower power consumption on BAN	Energy consumption	NA	Real time healthcare monitoring

Reference	Contribution category	Data type or data set	Main Objective of exploiting edge computing in healthcare IoT	Performance metric	ML-based approach	Use case
Muhammad et al. (2018a)	Framework, implementation	Various dataset	Enhance network Quality of Service, reduce latency	Latency reduction ratio, energy, classification accuracy, root mean square error, mean absolute error	Deep neural networks	Mobile healthcare framework for health monitoring, disease monitoring and remote super vision of surgery
Chen et al. (2018)	Framework, implementation	Medical history, physiological health risk level	Improve Quality of Experience in emergency case	Overall QoS	NA	Resource redistribution base on health risk level
Rahmani et al. (2018)	Framework, implementation	Physiological	Exploit higher-level service at the edge: local storage; real-time local data processing; embedded data mining; etc. improve energy efficiency, performance, interoperability, security and reliablity	Early warning score, latency ratio, compression ratio, latency	NA	Massive remote health monitoring system
Siddique et al. (2018)	Data compression	Physiological	Save energy	Energy consumption, SNR, PRD, PPR	Dictionary learning	Arrhythmia detection

Reference	Contribution category	Data type or data set	Main Objective of exploiting edge computing in healthcare IoT	Performance metric	ML-based approach	Use case
Priyadarshini et al. (2018)	Stress type classification, diabetes prediction, assessing hyper-tension's risk	Various	Reduce latency, increase throughput	Memory utilization, Classifcation accuracy	Deep neural network	Health monitoring and predict the wellness stats
Muhammad et al. (2018b)	Voice disorder detection, treatment framework	Saarbrucken Voice Disorder	Reduce bandwidth consumption	Bandwidth consumption, detection and classification accuracy	Deep learning	Voice disorder detection and classification
Sood and Mahajan (2018)	Disease classification	Physiological environmental location	Quick response, improve healthcare service	Classification time and classification accuracy	J48 decision tree	Diagnose and prevent the outbreak of Chikungunya
Pham et al. (2018)	Activity classification	Posture	Provide real-time data access, reduce the transmitted data, improve battery life	Energy consumption, classification accuracy	Gradient-boosting decision tree	Integrate contextual information for remote health monitoring
Liu et al. (2018a)	Food classification	UEC-256 Food 101	Recognize food for dietary assessment	Recognition accuracy, response time, energy consumption	CNN	Dietary assessment

Reference	Contribution category	Data type or data set	Main Objective of exploiting edge computing in healthcare IoT	Performance metric	ML-based approach	Use case
Dai et al. (2019)	Skin region classification	HAM10000	Reduce latency, improve privacy, lower cost	Classification accuracy	CNN	Detection and prevention of various diseases
Greco et al. (2019)	Framework, implementation	REALDISP	Quick response to anomalies	CPU usage, DB input and output throughput, DB CPU and memory usage, number of anomalies	Hierarchical Temporal Memory	Real time analysis of wearable sensor data streams
Pace et al. (2018)	Framework, implementation	Heart rate traces	Reduce the data traffic towards Internet	Transmitted traffic and processing time	NA	Heart rate monitoring to execute life critical service
Queralta et al. (2019)	Framework, implementation	MobiAct	Increase IoT device battery life, waive the need for sending raw data to cloud servers	Classification accuracy	LSTM RNN	Health monitoring
Hu et al. (2019)	Framework	Physiological	Minimize response delay	Disease's classification accuracy, response delay	Cascaded deep learning	Alarm system using chronic disease prediction
Kharel et al. (2019)	Architecture, implementation	Heart rate data	Minimize burden on the cloud, save energy	SNR, Path loss (dB)	NA	Healthcare remote monitoring

Reference	Contribution category	Data type or data set	Main Objective of exploiting edge computing in healthcare IoT	Performance metric	ML-based approach	Use case
Wu et al. (2019)	Implementation	Environmental, physiological	Process the raw sensor signals, monitoring the environmental and physiological data, trigger alert	NA	NA	Saftey and health monitoring
Uddin (2019)	Activity classification	Multiple healthcare sensors	Reduce healthcare cost, fast computation of experimental data	Classification accuracy	Recurrent neural network	Activity classification
Azimi et al. (2019)	Abnormaly detection and classification	Arrhythmia	Improve the availability, reduce response time	Classification accuracy, execution time, response time, transmission time	CNN	Detect and classify different abnormalities in ECG signals
Ram et al. (2019)	Activity classification	Physionet	Reduce large computing and communication overhead	Classification accuracy	SVM and random forest	Mobile health monitoring of human activity
Devarajan et al. (2019)	Risk-level diabetes prediction	UCI PAMAP2	Improve the bandwidth efficiency, energy, prediction accuracy, computational complexity and latency	Classification accuracy	J48 decision tree	Real time monitoring the blood glucose level

Reference	Contribution category	Data type or data set	Main Objective of exploiting edge computing in healthcare IoT	Performance metric	ML-based approach	Use case
Hayajneh et al. (2019)	Data protocol	Simulated	Channel assignment: avoid interference. Gateway association: improve the efficiency and security. Routing protocol: minimize delay and power at nodes	PDR, Delay	NA	Secure and efficient data delivery
Abdellatif et al. (2020)	Framework, implementation	Multimodal	Compress data, extract feature and detect seizure by deep learning based classification to boost the system performance, reduce the transferred data, preserve data privacy	Distortion ratio, battery level, classification accuracy	Deep learning, unsupervised learning, reinforcement learning	Realtime patient remote monitoring
Hegde et al. (2020)	Disease detection	Thermal video, visible light	Rapid response, separate interaction between patients and healthcare worker	Classification accuracy	Deep learning	Assist in the automatic triage of patients in emergency department
Abdel-Basset et al. (2020)	Type 2 diabete detection	Various biological	Improve response time and privacy	Execution time accuracy	Neutrosophic	Detecting and monitoring patients with type-2 diabetes

Table 1.4. *Edge computing in healthcare IoT*

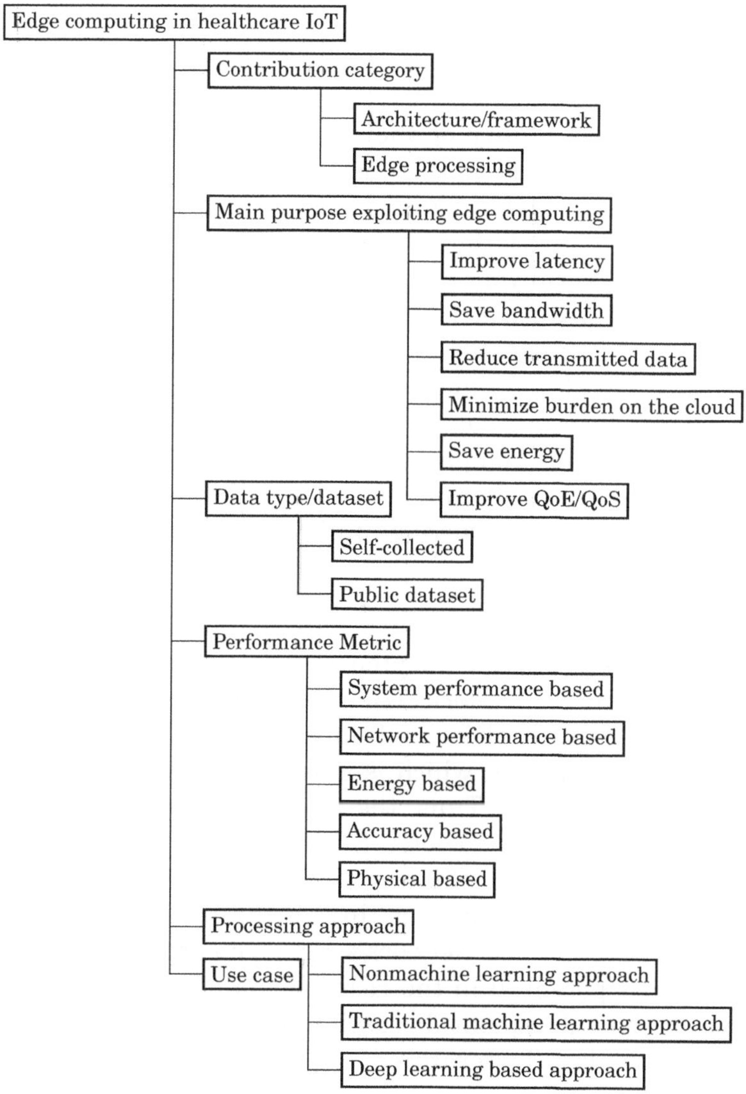

Figure 1.4. *Edge computing's taxonomy in Health-IoT*

The main reason for such challenges is the Internet's expansion on both horizontal and vertical scales to meet human demands. Besides, the heterogeneity and fragmentation in the field caused by the diversity of Internet software solutions and hardware equipment that have not been resolved effectively also escalate the situation. Not to mention, the compatibility of the existing infrastructure to the novel advancements in the field is also challenged by these complications. To handle such sophisticated state of the Internet, it is required the most resourceful organizations to manage the worldwide digital space. In this case, the companies (ISPs, MNOs, etc.) are the only feasible candidates. From end-user expectations, the Internet operation is supposed to be stable and work flawlessly. Hence, any disruption in Internet services could cause tremendous consequences to the companies' operations and revenues (Siddiqui 2021). A considerable expense will be saved if the operator can foresee and manage unfavorable events before their occurrences.

To deal with such uncertainty and complexity, a novel solution is introduced and adopted by the community, that is, software-defined network (Bannour et al. 2018). By decoupling the control plane and data plane, SDN provides a top-down network view, which is crucially vital in Internet management operations. Thus, the research community has proposed and adopted various SDN applications in the literature over the years. Sharing the equivalent expansion and diversity with the Internet, IoT is the next promising land for software-defined network's management capability. The paradigm's potential in the H-IoT domain remains to be explored. In the subsequent discussion, we uncover the SDN landscape in the H-IoT literature.

1.2.2.2. *Software-defined network's taxonomy in H-IoT*

Utilizing the same approach to the edge computing paradigm, we initiate the SDN's SOTA in H-IoT with architectural contributions.

Sinh et al. (2018) is one of the first works laying the foundation for this research direction. The authors addressed the role of SDN in IoT deployment, which is orchestrating the whole network. They also introduced an SDN-based architecture for IoT deployment.

Meanwhile, Baktir et al. (2018) presented a multi-tier computing and communication architecture for pervasive healthcare (including IoT) powered by SDN. The work aimed to investigate SDN's capability to handle the personalization of healthcare services and the delay sensitivity of continuous health assessment while exploiting the edge computing potential to deal with latency-intolerant services.

In addition to the SDN management enhancement to the H-IoT system, SDN also solidifies H-IoT infrastructure by enhancing its security. El Jaouhari and Bouabdallah (2018) studied the security mechanism including the confidentiality, integrity, authentication and access control of the H-IoT system. The current work was limited to the security management of a single infrastructure. However, the authors claimed to extend the extension in their future work. Nevertheless, the results indicated the necessity to improve the single-point controller performance to cope with the H-IoT security issues effectively. Taking the SDN bottleneck drawback into consideration while securing the H-IoT system, Aujla et al. (2019) proposed to move SDN from a cloud environment to an edge–cloud environment. The authors offered three main contributions to this work. First, they presented an offloading scheme aimed to support edge-cloud interactions. Their flow management scheme aimed to achieve the optimal flow path, reducing network delay even the number of healthcare devices and end-to-end distance increases. Finally, they also improve the system's security with a lattice-based cryptosystem. Taking a step further, Li et al. (2020) also adopted the edge-computing paradigm but for load balancing purposes. They also proposed a lightweight authentication scheme to enhance system security.

Besides proposing architectures that aim to improve security and centralized bottleneck aspects, other researchers also investigated various architectures to enhance general network performance in SDN-enabled H-IoT systems. In 2019, Cicioğlu and Çalhan (2019a) and Cicioğlu and Çalhan (2019b) presented their work about their SDN-based architecture and a novel energy-aware routing algorithm. The results confirmed their SDN-based structure meets the expected QoS and the routing approach is superior to the traditional routing approach in terms of energy consumption, throughput, successful transmission rate and delay. Afterward, Preveze et al. (2022) also studied the subject but focus their work on developing a novel adaptive switching technique. Via MATLAB simulation, they succeeded in increasing the throughput performance and promised to develop further the proposed system considering the energy factor.

Using SDN's orchestrating capability, Hasan et al. (2020b) proposed an SDWBAN framework to prioritize sensitive healthcare data over ordinary data. They employed an application classification algorithm and a modified sector-based distance (SBD) protocol version to implement data prioritization. Later, the authors developed a mathematical model to determine the optimal number of controllers for the proposed framework in Hasan et al. (2020a). The results show the latency and PDR performance improvement, which ensures the efficiency of their proposed SDN's administrative data scheme. Furthermore, Lakhan et al. (2022) also investigated this research direction but focused on a different aspect, the resource allocation problem. They modeled the SDN-enabled H-IoT system as a geographically distributed fog network, later formulating a multi-constraint optimization problem. A deep-learning Q-network-based resource allocation scheme was presented based on the prepared architecture and later proved its performance improvement. Misra et al. (2023) approached this research direction by offering on-the-fly analytics to the end devices. The SDN administration tasks in this work, such as reconfiguration and programmability, were considered at three stages: resource based, flow rule based, and analytics based. A large-scale simulation evaluation using Mininet highlighted the architecture efficiency by eliminating additional network delay. In the meantime, Sarkar et al. (2023) approached the problem under the server selection scope. They provided a framework taking advantage of game theory's approach to transfer information to the fog/cloud layer. In this work, they also utilized a three-layered CNN to detect patient's communicable illnesses. Jameel et al. (2020) also took advantage of SDN's orchestrating capability but focused on physical signal performance. The authors proposed a Q-learning-based optimization framework for mitigating the impact of interference in SDN-based backscatter heterogeneous networks. Their simulation results showed the increase in performance of the macro-cell users while maintaining a favorable SINR's level for backscatter users in the small cell.

Besides architectural contributions, various works concentrated on other specialized topics in SDN H-IoT-based systems. In the security aspect, in 2018, Meng et al. (2018) paved the way by introducing a Bayesian-based trust model for detecting malicious devices in SDN-based H-IoT. Their results demonstrated that the trust values of malicious devices using their proposed approach decreased faster than the SOTA, thus improving the identification stage. Later, in 2022, Radoglou-Grammatikis et al. (2022) presented a study about the IEC 60 870-5-104 protocol. Assessing this protocol, the authors provided a quantitative threat model that relied on attack defense trees and common vulnerability scoring system. They also presented an intrusion detection and prevention system derived from the proposed threat model. Various machine learning and deep learning techniques have been utilized to ensure their system's performance. Later, in 2023, Halman and Alenazi (2023) and Alzahrani and

Alenazi (2023) also focused on this direction and succeeded in increasing the SOTA successful detection rate impressively to more than 99%. In network performance, Misra et al. (2020) proposed a hybrid flow-rule replacement scheme that dynamically places or removes flow rules at the edge access point. The proposed method also considered users' criticality and mobility in healthcare scenarios. The achieved result confirmed significant latency, overhead and energy consumption reduction. From a scheduling perspective, Askari et al. (2021) proposed a two-tier mechanism for scheduling transmission from hubs to the APs. The proposed algorithm considered channel conditions, energy consumption and the highest priority to reduce the average delay and energy consumption and increase the effective throughput.

After the brief introduction to SDN's state of the art in H-IoT, we generalize research topics in the domain in Figure 1.5 and highlight chapter's content in Table 1.5.

1.3. Learned lessons and neglected opportunities

1.3.1. *Learned lessons*

Considering the presented state of the art, we structure our best understandings about explored topics of SDN and MEC in the H-IoT domain by Figure 1.6. From our point of view, we outline three major learned lessons from the literature's current state as follows.

First, the H-IoT system is usually formulated using the three-layer architecture: the Things, Processing and Communication layers. However, most researchers have been focusing on the processing tasks at the Things or Processing layers. Their investigations are the backbone for enabling future feasibility in this field.

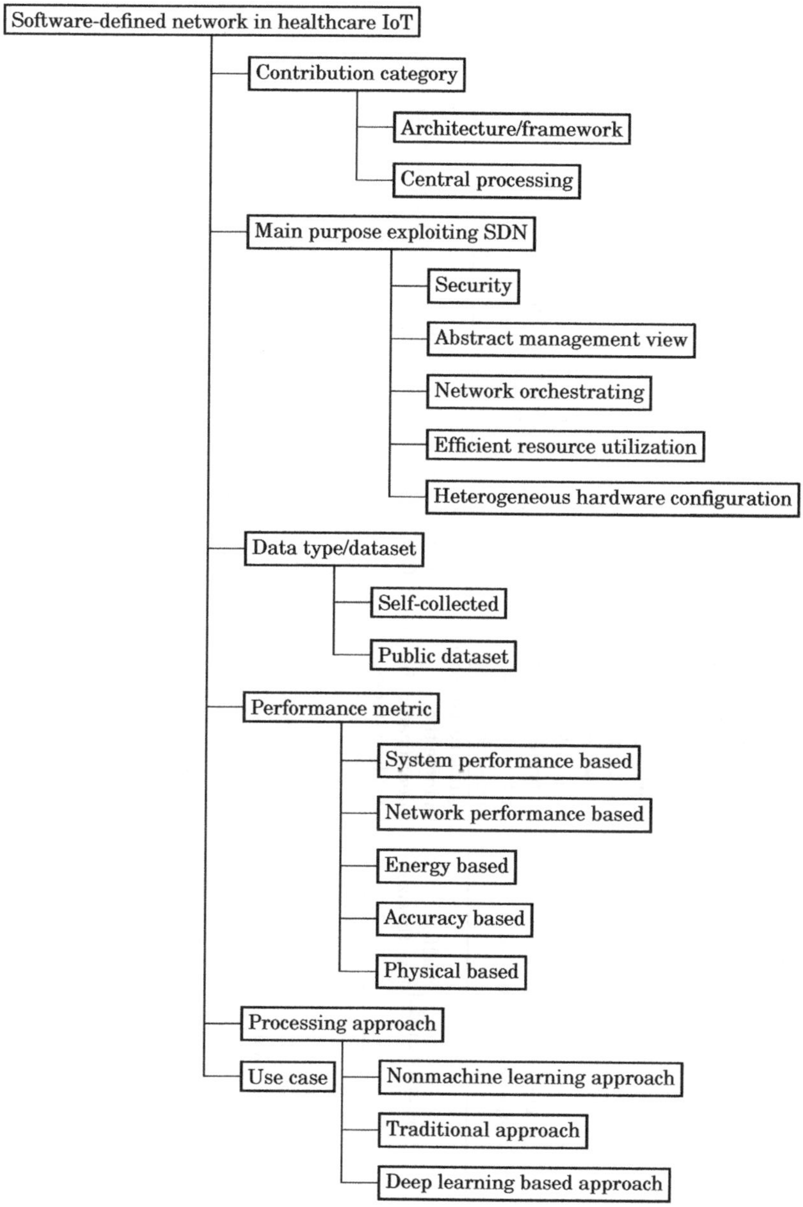

Figure 1.5. *Software-defined taxonomy in Health-IoT*

Reference	Contribution category	Data type or data set	Main Objective of exploiting software-defined networking in healthcare IoT	Performance metric	ML-based approach	Use case
El Jaouhari and Bouabdallah (2018)	Architecture	ZAP-based synthesized data	Solve securities issues (confidentiality, integrity, authentication, access control) in a smart infrastructure	RTT, CPU usage, request analysis, switch's network IO	Nonmachine learning	System security and monitoring
Sinh et al. (2018)	Architecture implementation	NA	Network orchestrating to build on-demand IoT service via network slicing	Service testing	Nonmachine learning	Network orchestrating
Meng et al. (2018)	Security model	Self-synthesized	Quick response to handle the identification in the insider attacks.	Trust value, CPU usage	Bayesian model	Intrusion detection
Baktir et al. (2018)	Architecture, fall risk assessment	Self - synthesized	Efficient manage the resource allocations and facilitate personalized healthcare services	Service delay, CPU usage, classification accuracy, decision boundary	SVM and random forest	Fall risk assessment
Aujla et al. (2019)	Framework	Self-synthesized	Real-time analysis with seamless and secure transmission, ensure Quality of Service	Offloading rate, latency, complexity, network delay, number of handover	Nonmachine learning	QoS assurance

Reference	**Contribution category**	**Data type or data set**	**Main Objective of exploiting software-defined networking in healthcare IoT**	**Performance metric**	**ML-based approach**	**Use case**
Cicioğlu and Çalhan (2019b)	Routing algorithm	Self-synthesized	Centralized control to integrate and manage network infrastructures through a standard interface	Throughput, successful transmission rate, end-to-end delay, energy consumption	Nonmachine learning	Inter-WBAN communication
Cicioğlu and Çalhan (2019a)	Architecture, Interface protocol	Self-synthesized	Complex network infrastructure management	Throughput, delay, PDR, average energy consumption, BER	Nonmachine learning	Handling heterogeneous WBAN structures
Hasan et al. (2020b)	Framework	Simulated data (Castalia)	Provide administrative control for WBAN	PDR, CDF, latency	Nonmachine learning	Real time patient monitoring
Li et al. (2020)	Framework	Simulated data	Efficient resource utilization	Delay, throughput, control overhead, average response time, PDR	Nonmachine learning	Securing system and and effective resource utilization

Reference	Contribution category	Data type or data set	Main Objective of exploiting software-defined networking in healthcare IoT	Performance metric	ML-based approach	Use case
Hasan et al. (2020a)	Network optimization	Simulated data (Castalia)	Abstract view to manage network and efficient resource utilization	PDR, delay	Nonmachine learning	Exploring mathematical relationship
Misra et al. (2020)	Person's criticality trajectory prediction	Heart Disease UCI	Heterogenous hardware configuration, better control in traffic forwarding, channel allocation	Accuracy, latency, energy, overhead	Decision tree, Naive Bayes	Prioritize healthcare traffic
Jameel et al. (2020)	Q-learning model	Time-series heart rate data	Heterogenous hardware configuration	SINR	Reinforcement learning	Backscatter communication in healthcare
Askari et al. (2021)	Multi-access schedule algorithm	Simulated data	Abstract view to manage network and efficient resource utilization	Effective throughput, average energy consumption, average delay	Deep neural networks	Real-time healthcare monitoring

Reference	Contribution category	Data type or data set	Main Objective of exploiting software-defined networking in healthcare IoT	Performance metric	ML-based approach	Use case
Radoglou-Grammatikis et al (2022)	Quantitative threat model	Synthesized IEC 60 870-5-104 dataset	SDN's attack mitigation mechanism	Detection performance, classification accuracy, mitigation performance	Multi armed bandit, classification regresion decision tree, reinforcement learning with Thomson sampling	Secure healthcare infrastructure
Alzahrani and Alenazi (2023)	Attack detection	MCAD-SDN	Ensure efficient resource utilization, security, optimal network control and management	Detection classification accuracy	KNN, decision tree, random forest, naive bayes, logistic regresion, ada boost, extreme gradient boosting	Secure healthcare infrastructure
Preveze et al. (2022)	Adaptive switching algorithm	MATLAB simulation	Manage complex healthcare IoT networks	Throughput	Nonmachine learning	Real-time health monitoring
Lakhan et al. (2022)	Architecture, lightweight Q-model-based policies	Simulation	Handling challenged resource allocation in mobility and distributed dynamic fog networks	Resource utilization, latency, energy consumption	Deep reinforcement learning	Resource allocation

Reference	**Contribution category**	**Data type or data set**	**Main Objective of exploiting software-defined networking in healthcare IoT**	**Performance metric**	**ML-based approach**	**Use case**
Halman and Alenazi (2023)	Attack detection	MCAD-SDN	Ensure efficient resource utilization, security, optimal network control and management	Detection classification accuracy	KNN, decision tree, random forest, naive bayes, logistic regresion, ada boost, extreme gradient boosting	Secure healthcare infrastructure
Sarkar et al. (2023)	Communicable illness detection	Simulation	Reduce latency, increase total network longevity	Delay, energy consumption	CNN	Illness detection
Misra et al. (2023)	Architecture	Simulation	Identification, dynamic management of services, and fast analysis	Response time, latency, PDR, throughput	Nonmachine learning	Remote healthcare monitoring

Table 1.5. *Software-defined networking in healthcare IoT*

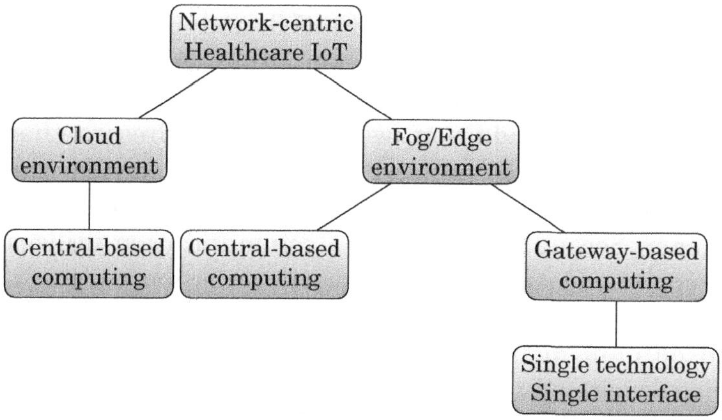

Figure 1.6. *Explored research topic generalization in network-centric healthcare IoT*

Furthermore, most research is being conducted with a few presumptions: the end user gateway and the designated central server are identified and expected to be kept unchanged. At the same time, the middleware network infrastructure is *managed by MNOs and ISPs* and *exclusively* depend on them to ensure *a stable channel*. Despite the development of various communication technologies extending both the range and bandwidth for the transmission channel, these assumptions could only be satisfied under the following conditions. Both client–server endpoints are *static and utilize an unyielding* network infrastructure for their interactions. In other words, the communication infrastructure must be stable throughout the H-IoT transmission process. However, these criteria are hard to convince as the only typical scenario in our studying domain, since one endpoint mainly is the human body, and its mobility factor requires considerate attention. Besides the end-users mobility, among various challenging factors, such as the high-density deployment of advanced middleware network infrastructure (5G and 6G) and the dynamicity in complex central computing environments, etc., the expectation about the above network conditions seems unrealistic. In addition, we have also addressed that the 5G novel concept, network slicing, has opened the opportunities to hand over the network management to the customer. Second, current H-IoT communication utilizes only *one communication interface* for transmission purposes. Nevertheless, no study or countermeasure in the domain is proposed or implemented to allow for a seamless network experience despite any unpredictable obstacle that may cause a function disruption to the only interface. Considering the operational and environmental uncertainty, especially in the context of this work, that is, healthcare emergencies, it is also challenging to guarantee the reliability of the client-side transmission in such conditions.

Lastly, the H-IoT system formulation requires numerous design decisions depending on each use case. After the system design process, the technology combinations in the three-layer architecture must be chosen. The community has offered various solutions in each subdomain in the physical and virtual layers. However, this diversity without a common standard also causes heterogeneity and fragmentation issues to develop and deploy the decided solution.

A noteworthy example of this situation is the diversity in communication technology in the Communication layer (Tables 1.2 and 1.3). Thus, these problems have been challenging academia and the industry for decades and have not been resolved effectively yet.

From the deriving knowledge, we explain our critical thinking process and the reason for our research decisions in this work in the subsequent discussion.

1.3.2. *Neglected opportunities*

Exploiting the network landscape preview and our learned lesson about the SOTA of SDN and MEC paradigm in the H-IoT domain, we list the neglected opportunities awaiting to be discovered by the community as follows:

– The opportunity to manage the network infrastructure has yet to receive its deserved attention from the eHealth researchers despite its promising impact.

– Since most of the research is based on stable network and infrastructure conditions, more foundation investigations should be conducted toward flexible and dynamic network infrastructure in the H-IoT domain. The usage of multi-technology and multi-radio utilization and their potential in the domain have yet to be fully explored, despite being briefly mentioned in some work. A unified solution to the domain's heterogeneity and fragmentation issues has yet to be proposed in the literature.

Considering all of the insights above and the criticality of delivering vital healthcare data in real-time in H-IoT urgent situations, we recognize the missing gap between stable and static and flexible and dynamic network infrastructure in the H-IoT domain. Besides, the open challenges in dealing with heterogeneity and fragmentation in the literature are awaiting to be resolved. These facts motivate our research toward providing the unveiling solutions for the H-IoT domain in such directions. While proposing solutions for the heterogeneity and fragmentation issues in the H-IoT domain, we also harness the undiscovered prospect of utilizing multiple communication interfaces simultaneously for the best network performance and cutting-edge transport protocols to maintain a seamless network experience for the terminal node, especially in the 5G era's context. Despite focusing on the most challenging and uncertain network conditions in the H-IoT branch, our research prospects are unlimited and applicable to the general IoT context or other domains. Figure 1.7 depicts the book's research directions and our contributions' positions in the innovated literature.

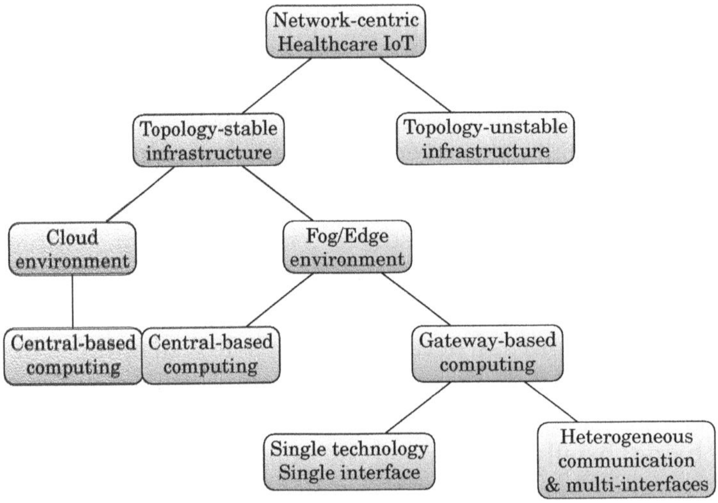

Figure 1.7. *Proposed research directions in network-centric healthcare IoT. For a color version of this figure, see www.iste.co.uk/chu/softwaredefined.zip*

1.4. Conclusion

This chapter has presented an overview of the H-IoT domain and its key enablers. We also create taxonomies from the studied work, which later help us extract the learned lessons and neglected opportunities in the domain. From the extracted insights, we navigate our research to focus on the undiscovered directions that take healthcare emergencies into prioritized consideration. However, thanks to their genericity, the prospect of the proposed solutions is applicable to the general IoT context or other domains. The following chapters will describe our proposed solutions and their prospective impacts in more detail. The next chapter presents our first contribution of software-defined network, a novel network paradigm.

Chapter 2

A Novel Network Infrastructure Concept

Software defined networks and the maturing of the Internet.

Nick McKeown

This chapter presents the first contribution in the H-IoT domain, a cutting-edge network paradigm that unlocks current *network infrastructure's* understanding.

2.1. Introduction

As mentioned earlier, delivering vital healthcare data in real time is an essential factor behind the success of H-IoT applications. Thus, this particular characteristic demands the most attention from the communications infrastructure, with consideration toward an architecture dedicated to this type of application. Hence, this chapter focuses on software-defined infrastructure (SDI), a novel paradigm that aims to advance the H-IoT's state of the art while prioritizing support for the most challenging scenarios in the domain, that is, healthcare

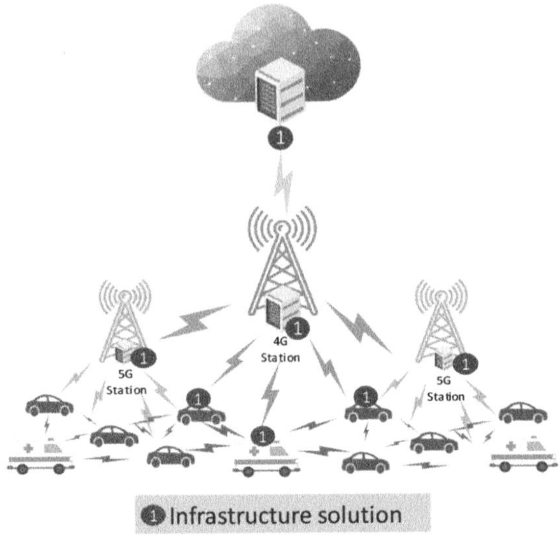

Figure 2.1. *First contribution's prospect in the overall framework. For a color version of this figure, see www.iste.co.uk/chu/softwaredefined.zip*

emergencies. Figure 2.1 highlights this chapter's influence on the framework.

Not until 5G's existence, the researchers in the eHealth domain commonly ignored the managing possibility of the backbone network infrastructure. The reasons for this unwilling dependence are the operational expense barrier and the exclusive infrastructure management of Internet service providers (ISPs), mobile network operators (MNOs), etc. However, 5G provides a solution for the researchers to overcome these limitations by employing network function virtualization (NFV) and network slicing concepts (Barakabitze et al. 2020). Specifically, these advanced features allow third-party network management and operation on a self-configured network overlay on top of a predeployed 5G network infrastructure. The contemporary vision of third-party network operation presents the community with a remarkable concept transformation. Thus, this phenomenon also steers the community to revise existing research, especially in the eHealth domain, to employ the novel paradigm. However, taking a closer inspection, the primary fuel behind NFV and network slicing to enable such innovative network prospects is a well-known paradigm, that is, software-defined network (SDN) (Tong et al. 2022). Notably, SDN's network model is composed of two components: the central intelligence will manage all other components, and the simple switches will handle the network forwarding process. Hence, SDN functionality in each 5G infrastructure equipment is a must to support the mentioned prospects. However, a global hardware replacement campaign is impractical, considering the current state of the Internet. Hence, the hybrid approach (Khorsandroo et al. 2021), which takes advantage of both SDN-enabled and traditional network devices, is exploited to provide the SDN-enabled capability to existing infrastructure. The traditional network devices that maintain Internet services usually are switches, routers, etc. The common features among all categories is their static and inflexible deployment method (Fogli et al. 2022).

Meanwhile, numerous developments toward a more flexible network infrastructure also received significant attention from the community. Thanks to wireless technology's advancements (5G and 6G) (Xu et al. 2021; Salameh and El Tarhuni 2022) and the shifting trend to embed the communication component in mobile devices, such as mobile phones, vehicles and unmanned aerial vehicles (UAV), the network architecture similarly evolves itself. While the deployment of access point coverage has become significantly more feasible, any entity that possesses the requisite communication technologies is capable of acting as a network node and provides various advanced network applications (Liu et al. 2022). Specifically, this entity could deploy network coverage and forwarding capabilities as well as consume network coverage from other nodes as an actual network node. To achieve convergence with the evolution of modern network architecture, the SDN paradigm requires at least equivalent levels of exertion and advancement. Hence, software-defined unmanned aerial vehicle (SDUAV) (Alharthi et al. 2019) and software-defined vehicular network (SDVN) (Xu et al. 2022), the two most modern SDN architectures, are born to adopt software-defined paradigm in such elastic circumstance. These candidates are also advantageous in critical situations requiring large-scale coverage deployments such as disaster scenarios (Lebre et al. 2020), crisis circumstances (Munawar et al. 2021), medical emergencies, etc.

However, the current approach of the community has revealed some inherent constraints. By applying the software-defined paradigm to some specific contexts (Alharthi et al. 2019; Xu et al. 2022; Lebre et al. 2020; Munawar et al. 2021), this approach endows the research community with the direction to define the technical specifications for some predetermined operating environments. Nevertheless, following this approach may help the researchers to narrow their working scopes but also establish an apparent drawback, limiting the SDN paradigm management's generalization. Despite the existence of a plethora of network environments, such as wireless body area network (Singh and Prasad 2022) or under water network (Unal et al. 2022), the number of research papers toward SDN's adoption in those environments is the minority compared to vehicular network and UAV network. Hence, the literature lacks a unified mechanism to cope with the heterogeneity and adaptation in diverse SDN environments in hardware and software aspects.

Due to this fact and the vast development of modern network infrastructure, the modern SDN's concept requires a major revision considering neglected potential opportunities not only in the existing context but also in anticipated circumstances (Fogli et al. 2022). In this contribution, we attempt to deal with this research gap by extending an existing paradigm, SDI (Li et al. 2019b), used for virtual network infrastructure within a static network infrastructure. The extended paradigm not only enlarges 5G's cutting-edge network perspective but also ensures the simplicity of the software-defined paradigm implementation in any existing or prospective environments. Following our proposed idea, we aim to inspire the community toward a shared goal by fully exploiting SDN capability in this study.

As a result, we present in this chapter a unified, flexible, dynamic and automatable network infrastructure in software and hardware aspects. Among various applicable use cases in the eHealth domain, we target the most uncertain and challenging network environment to highlight SDI's advantages. Hence, we selected the delivery of vital healthcare data in emergency scenarios for evaluating SDI's improvements. Our contributions in this work are encapsulated as follows:

– Introducing a novel point of view that extends the "software-defined infrastructure" paradigm, later enlarging 5G's network perspective for the next Internet generation.

– Formulating a unified, flexible, dynamic and automatable network infrastructure from the extended paradigm.

– Proof-of-concept SDI's evaluation framework toward further investigation from the community.

– Four SDI-empowered extensive evaluation approaches in eHealth's vital data transmission use cases (opportunistic network enhancement, self-assisted coverage deployment and data transportation, connection recovery, and priority orchestration in high-density network) where the researchers could take over the network infrastructure management.

The rest of this chapter is structured as follows. Section 2.2 provides a brief understanding of the required background and existing work. We describe our facilitated network infrastructure and its potential applications in section 2.3. Section 2.4 provides a formalization of the unified functional model. The experiment discussions are divided into two sections: experiment description in section 2.5 and result analysis in section 2.6. Finally, section 2.7 presents the conclusion of this chapter.

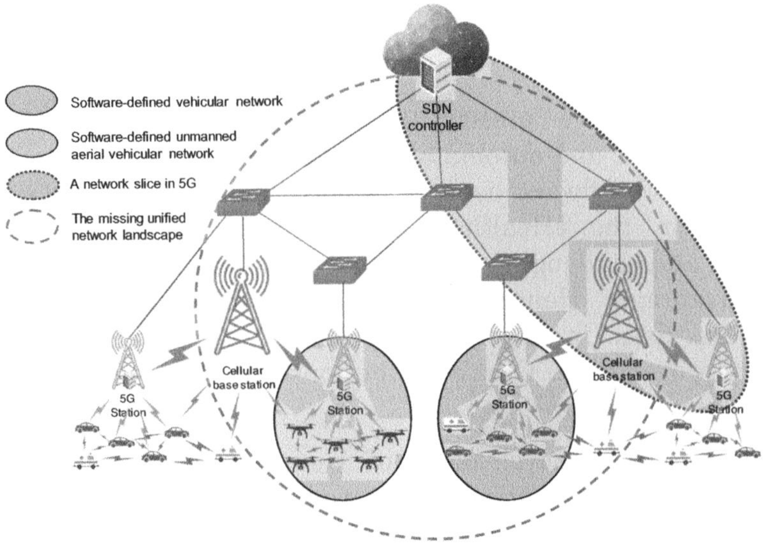

Figure 2.2. *General network context. For a color version of this figure, see www.iste.co.uk/chu/softwaredefined.zip*

2.2. Related work

The fundamental topics of the SDI extension will be explored in this section. The first discussion defines the SDN paradigm. Next, we presents the origin of the most recent network perspective in the 5G's introduction. In the third place, the history of the "software-defined infrastructure" terminology in the literature will be specified. We also address the perspective fragmentation in the general network context to highlight the missing network perspective in the literature. Finally, the definitions of each suggested potential hardware block for the imminent SDI paradigm will be covered afterward.

2.2.1. *Software-defined network*

SDN, the crucial concept that inspired our proposal, is the paradigm that introduced the novel approach to handling network management. By using physically distributed simple data planes and a central logic controller (centralized or physically distributed), the SDN paradigm allows simplicity as well as a global solution in managing diverse network hardware environments. This approach not only provides the MNOs, ISPs, etc., with a more efficient way to operate their network infrastructure but also enables a brand new perspective about the computer network, which leads to numerous novel research directions such as advanced SDN-enabled applications for large-scale network (Tong et al. 2022) or SDN for wireless environment (Fogli et al. 2022).

2.2.2. *5G novel network perspective*

The SDN's two-plane network model has empowered the network infrastructure management aspect and enabled two 5G's advanced features: network function virtualization and network slicing (Barakabitze et al. 2020). First, NFV offers the capability to manage a shared physical infrastructure to deploy a virtual network infrastructure. On top of such sharing virtual network infrastructure, various network overlays could be created and managed by SDN. Each overlay is considered a network slice with on-demand characteristics. Hence, the act of SDN to optimize the overlay's service specifications refers to network slicing terminology. The network slicing unique function could allow third-party actors to influence the deployed network infrastructure, which used to be exclusively managed by MNOs or ISPs. The novel network perspective profoundly changed the community network perception, which commonly used to neglect the backbone network infrastructure management of the MNOs and ISPs. In the current 5G implementation, network slicing indicates three first specifications: enhanced mobile broadband, massive machine type communication, and ultra reliable and low latency. However, in the recent study, the future of embedding dedicated prioritized slices or strict requirement use cases is feasible (Subedi et al. 2021), which is essential in the eHealth domain regarding the context.

2.2.3. *Software-defined infrastructure*

Also inspired by SDN paradigm, SDI is constructed for the first time in the literature in the work of Lin et al. (2014). In this paper, the paradigm refers to the virtual network infrastructure managed by a software-defined controller, which operates on top of an existing physical multi-tier cloud infrastructure.

Subsequently, as an ongoing effort, Kang et al. (2014) presented an implementation of the SDI concept in a tangible experimental setup named smart applications on virtual infrastructure (SAVI) in 2014. Until recently, SDI's usage in the literature (Parsaeefard et al. 2019; Li et al. 2019b; Huang et al. 2019; Beigi-Mohammadi et al. 2020) is still limited within the same defined context. Regardless of the community's modest vision, each piece of the paradigm still vastly bloomed in the literature.

Figure 2.2 represents the current general network context in addition to the latest network advancements. The dominance of the cloud environment in the Internet's backbone operations is evident. To address high latency and scalability challenges, computing resources are brought closer to end-users by adopting multi-access edge computing in vehicular and UAV environments. In addition, recent advancements in communication technologies, specifically giga hertz in 5G and followed by tera hertz in 6G, have facilitated the deployment of the base station's new generation. Besides SDN-enabled base stations, SDN controllers and the existing infrastructure are also essential to support 5G's NFV and network slicing functionality. Instead of substituting the existing backbone infrastructure entirely, SDN's equipment strives to provide the best solution based on users' demands by collaborating with the existing infrastructure. The figure also illustrates the fragmented scope of each outstanding SDN-inspired paradigm in the literature: network slice (in 5G infrastructure), software-defined vehicular network and software-defined unmanned aerial vehicular network. While 5G's network slicing and NFV potential is limited in the static network architecture, the unified collaborative prospect between SDUAV, SDVN and other potential SDN-inspired paradigms needs to be determined. In conclusion, each paradigm has operated independently and is limited in its fixed technical scopes. Clearly, the SDI's latest meaning has failed to fully capture the modern network concept because of the extensive developments in the field. The following discussions will further study each existing hardware block before its fusion in the new SDI paradigm prospect, and later described in section 2.3.

2.2.4. *Software-defined vehicular network*

The first SDI's hardware block is the software-defined vehicular ad hoc network (Islam et al. 2021). A *Vehicular Ad hoc NETwork* (VANET) is created by adopting an ad hoc mobile network in the vehicular domain. This paradigm promises various benefits for passengers and drivers, such as safety, convenience and infotainment (Mekki et al. 2021). However, because of the vehicular environment's natural characteristics, several fundamental challenges, such as unreliable connectivity, seamless handover and heterogeneous network environment, must be considered. By providing a central view of the network, this distinguishing feature of the SDN paradigm is a crucial solution for existing problems of VANET (Islam et al. 2021) to allow for advanced future applications, such as improving road safety and reducing carbon emissions (Benaissa et al. 2021). Software-defined vehicular ad hoc network is facilitated by leveraging the software-defined paradigm in VANET.

Regarding the VANET extension in the real world, vehicle-to-everything (V2X) is the proposed technology that allows for establishing coverage in a vehicular environment. Campolo et al. (2018) proposed a practical slicing framework and related mechanism to support V2X applications in 5G network infrastructure, especially in roaming scenarios.

On the other hand, the idea of exploiting the ability to deploy network infrastructure by vehicles to allow seamless handover in vehicular environments has been initially mentioned in Silva et al. (2021). In this paper, the authors proposed to use a vehicle (onboard unit) to extend the network coverage for end-users. However, using this approach may not achieve the full potential of the software-defined controller in their proposed system. Since the controller is not yet able to improve the vehicle's network management and influence the vehicle's position and trajectory to provide the best performance for other actors.

2.2.5. *Software-defined unmanned aerial vehicular network*

The second crucial SDI hardware block in the SDI paradigm is SDUAV. Similar to software-defined VANET, the approach for inventing "software-defined UAV" is based on adopting the software-defined paradigm in the UAV environment. Sharing equivalent characteristics with VANET, such as mobility and unreliable connectivity, UAV networks equivalently possessed similar potential problems. However, the UAV operating environment is much more flexible than VANET, making it feasible for surface-based disaster or no-road scenarios. Moreover, the UAV's primary advantage also constitutes a significant limitation. While its aerial operation enables versatile deployment, it simultaneously imposes constraints due to limited power capacity inherent to flight-based systems.

Zhang et al. (2018) took advantage of the "software-defined" paradigm to propose a UAV backbone framework and a load-balancing algorithm to maintain a desirable network service. In this work, The SDN and UAV controllers could perform network and location management. However, since the view is focused on specific hardware, the proposal did not aim to unify the "software-defined" paradigm in a much more general manner.

In 2018, Dai et al. (2022) presented a proposal to take advantage of both UAV and base stations for mobile edge computing deployment. In the meantime, Abkenar et al. (2022) also presented a UAV-assisted fog-IoT framework for disaster management to implement their energy-efficient algorithms.

A year later, Alharthi et al. (2019) proposed an architecture for a software-defined UAV network. At the same time, Mozaffari et al. (2019) presented a comprehensive survey about the potential of UAVs for wireless networks. A holistic overview of the UAV's capabilities was given. However, the point of view was still limited to UAV-based technologies.

2.2.6. *Other potential software-defined hardware blocks*

As stated in the introductory section, our objective is to formulate a unified, flexible, dynamic and automatable network infrastructure. As such, prospective hardware blocks should be acceptable within the renovated SDI ecosystem's context. Hence, several promising blocks that may be suitable for SDI proposal will also be listed in this discussion, such as software-defined Internet of Things (SDIoT) (Ja'afreh et al. 2022), and software-defined cellular network (SDCN) (Vassilakis et al. 2017) and software-defined radio (SDR) (Kafetzis et al. 2022).

Software-defined Internet of Things is the combination of software-defined paradigm with Internet-of-Things, which aims to use an SDN controller to orchestrate network management in an Internet-of-Things environment, hence it is a feasible candidate for SDI ecosystem.

Software-defined cellular network is a paradigm fabricated by the emerging software-defined paradigm in the cellular network environment, which aims to enhance the network management of cellular base station infrastructure.

Another potential SDI candidate is the software-defined pedestrian network. The combination of software-defined networks and opportunistic pedestrian networks inspires this novel paradigm. A well-known opportunistic-oriented network example is pocket switch network (PSN) (Hui et al. 2005), which makes use of human mobility and local forwarding to transfer the data. The main idea of this paradigm is that information could be stored and passed within the devices as long as proper contact is met using wireless links. Note that this network paradigm's transmission progress depends on the ad hoc movement of the human body. Hence, the required information might not arrive at the destination after all.

2.3. The evolution of SDI

2.3.1. *Research gap*

After a brief understanding of existing technologies and essential and promising SDI hardware blocks in the literature, we again emphasize that all of these technologies are tightened up with a specific kind of hardware and a specific deployment environment. To build a coherent software-defined ecosystem, a unified solution is mandatory to emerge the "software-defined" paradigm in any network device as a solution to cope with heterogeneous hardware and the various physical operating environments of the existing devices or any future equipment. Thus, the following discussion presents the extension of SDI's definition to achieve the described prospect.

2.3.2. *SDI extension*

Prior to the existing scope of SDI in the literature, SDI was employed as a virtual computer network infrastructure fabricated on top of a high-performance and wired-up hardware computer system.

Nowadays, such complex, static and advanced network architecture, whose expanding process requires substantial endeavor (investing in new equipment, adamant and consuming installation and deconstruction effort, high maintenance expense, etc.), is essential to maintain the backbone of the Internet and crucial IT operational activities in the world. However, thanks to the advancements in wireless technologies, several versatile solutions to deploy network functionalities have been implemented, such as vehicle base stations, UAV base stations and personal mobile phone hotspots. Therefore, the previous SDI's context needs to be updated and requires a revisitation.

Since the term "software-defined infrastructure" has already been coined but has been limited to a specific context, we propose redefining this terminology into a more comprehensive and general paradigm. The definition extension no longer limits the terminology to mean software-defined virtual infrastructure but includes actual physical hardware and hardware deployment aspects. By articulating this idea, software-defined infrastructure in this work represents a network ecosystem composed of static network hardware infrastructure and a central controller's capability to deploy extensible hardware network infrastructure to provide flexible network coverage. Besides, the deployed infrastructure's network functionalities will also be managed by that genuine controller and/or capable of delegating controlling authority depending on the circumstances. In the reverse direction, any SDI-equipped device can also be involved or participate in any existing SDI in the area.

Following the renovated paradigm, the subsequent network slicing perspective in 5G could be described as follows. Capable of controlling both physical and virtual aspects of the network, the controller could employ the entire network landscape in its managed area: the existing static infrastructure, the flexible self-deployed infrastructure and the third-party infrastructure participants. Hence, the collaborative prospects among those entities are enabled and unified. The SDI controller could create and manage additional fully customized network slices in such a dynamic, heterogeneous, but unified network ecosystem. Thus, the presented landscape has enlarged 5G's prospect in flexible infrastructure and its vision clearly beyond 5Gs or any other software-defined inspired paradigms. As a result, the definition's limitation in the software-defined infrastructure paradigm is eliminated, which encourages its prospective unconstrained essence. More importantly, our terminology extension has elaborated the cutting-edge 5G vision and prepared the community's next step for the following technology generation. Afterward, we define a literature network landscape with the extended SDI definition.

2.3.3. *Proposal of our architecture*

Unlike previous works that refer to SDI as a virtual infrastructure on a rigid and unyielding network infrastructure, novel SDI architecture could be formulated using both virtual and physical components. Hence, the novel paradigm provides a unified network landscape encompassing various cutting-edge SDN-inspired paradigms (including SDVN, SDUAV and 5G network slicing) and prospective paradigms to a shared vision. This work focuses on the SDI's physical aspect deployment because of the scientific connection between such aspects and our contribution in this work. By describing the scientific relationship between physical hardware deployment and our contribution in this work, we refer to the SDI's capability to deploy extra network components (vehicle, UAV, etc.) for aiding healthcare transport communication of the extended SDI paradigm.

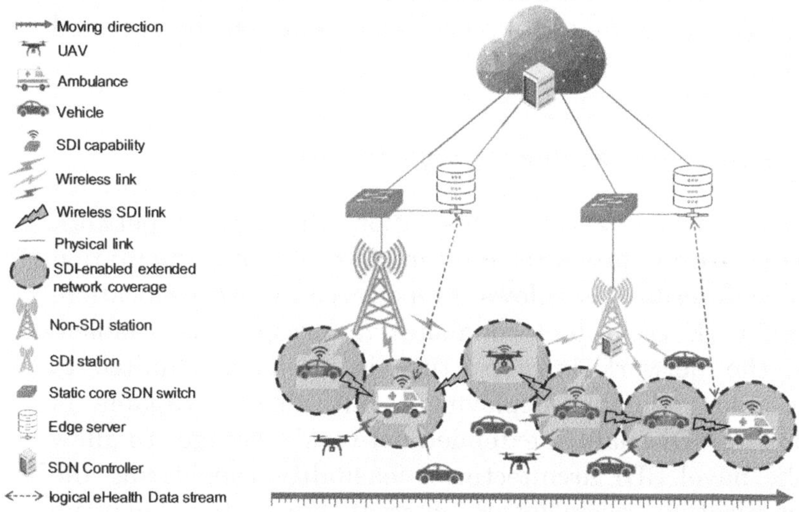

Figure 2.3. *SDI-empowered modern network infrastructure. For a color version of this figure, see www.iste.co.uk/chu/softwaredefined.zip*

Figure 2.3 describes the novel SDI's architecture after its definition extension. Each icon illustration is explained as follows. Two arrows are used for different purposes. The uni-direction blue arrow illustrates the moving directions of the follow-up physical vehicles: UAV, ambulance and vehicle. On the other hand, the bidirectional red arrow indicates the logical eHealth data stream between the ambulance and the edge server. Base stations provide large-scale coverage for the infrastructure; however, two base station symbols are used to characterize the ancient base station and the SDI-enabled base station. The network links between each component could be physical (thin blue line) or wireless links (lightning symbol). The SDI links are highlighted using different lightning symbols and a coherent theme with SDI-enabled extended network coverage. Cloud computing environment, SDN controller and static core SDN switches, and edge computing servers are deployed to enable the most advanced network function for the infrastructure. Nevertheless, an open-flow enabled switch combined with a wireless signal icon is used to embed the enabling SDI-capability on the specified vehicles.

2.3.4. *Suggested technology adjustments*

After the architectural illustration, the modern network infrastructure's state after the complete SDI implementation could be described as follows. Empowered by SDI unification, the network ecosystem collaborative prospect is enabled. First, the network infrastructure is no longer limited to static and inflexible deployment but also is capable of employing SDI-enabled extended network coverage. To allow for the novel SDI architecture's feasibility, considering the communication technologies' current state, the following modifications in the state of the art are recommended to be implemented to achieved the proposed prospect.

The foremost challenge is a generic technology stack (both hardware and software) to allow any equipped device to become SDI enabled. In such a technology stack, the network stack in each operating environment requires the most updates. Since each device that participates in the architecture could be managed by the central controller, it should be able to forward traffic on behalf of its neighbors. In other words, each device should also be SDN manageable. Any actors with network capability, whether mobile or not, public or private, could participate in the system to augment existing infrastructure. The infrastructure enhancements that could benefit the entire ecosystem could be coverage flexibility, performance adaptability, etc., for the other participants in their required use cases.

Along with the operating system level update, a unified control protocol is essential to allow the maximum controller's side controllability and resolve the heterogeneous technology problem. Hence, we propose to choose the current well-known SDN protocol, OpenFlow, as the generic communication protocol for the SDI ecosystem. We also suggest extending this protocol to support the node's location information collection and management for flexible hardware deployment. Figure 2.4 indicates a simple integrating concept of SDI functional block in a potential SDI hardware block. Depending on the specific situations, the SDI function could be provided as an independent software package to integrate with existing network hardware or a unified solution embedded directly into the nonnetwork equipment. However, the core SDI functional block will have four main modules to suit its purpose. The three first modules required for the ordinary network functions in a potential software-defined infrastructure hardware block are the wireless coverage access control module (manage the wireless coverage accessibility), the forwarding tables (network routing functions) and the network interfaces (abstract layer of physical network devices). However, the most important module is the SDI-enabled module, whose job is to handle the SDI controller's commands and manage other modules to follow the received instructions. Without the SDI controller's reachability, this module will manage the sensing of existing SDI-enabled devices in the area to coordinate and maximize the SDI collaborations. Notably, the SDI functional block communicates with the location module for location descriptive as input for prospectively related computation.

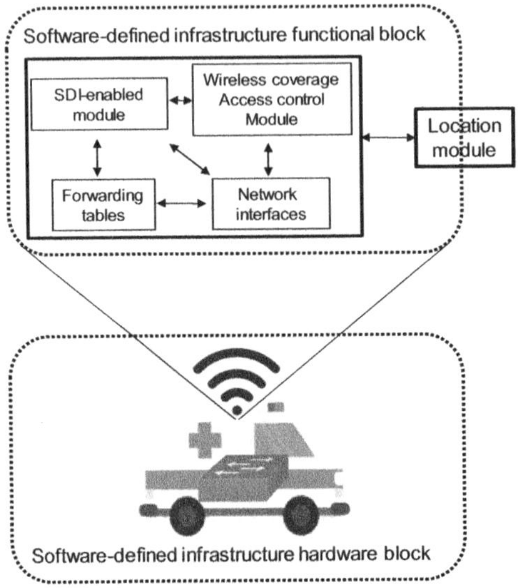

Figure 2.4. *SDI's practical integrating prospect*

2.3.5. *Prospective interaction within the SDI ecosystem*

The following discussion will describe the interaction between each hardware block in the SDI ecosystem after all of the proposed changes have been implemented.

Primarily, the requisite handling capacity of the central controller is mandatory. The whole SDI ecosystem will be managed by an SDI controller that can adapt to any existing or future technology. While managing existing network infrastructure to provide general network functionalities for the population, the central software-defined controller also monitors the current state of the network. Depending on each circumstance, the controller will control existing hardware or deploy further hardware, such as a vehicle and UAV, to enhance the network quality. The controller could also involve outside actors to recover, improve or even establish the network coverage and capability for desired locations.

SDVN and SDUAV terminologies are crucial to build the initial platform for this proposal. First, SDVN will empower SDI with automotive capability, with its many advantages in terms of power and coverage, and will mainly take on the role of deploying backbone infrastructure. Second, SDUAV, with its air deployment flexibility, will dynamically enhance the network's performance for the required target area.

Not only two solid candidates, but the other potential SDI hardware blocks could also be considered to strengthen the proposal. While SDCN could collaborate with SDVN and SDUAV to provide broad-scale coverage, SDIoT and PSN could provide more tailored or circumstantial coverage.

Besides, inspired by pocket switch network (PSN), there may be an emerging software-defined paradigm for pedestrians. In this promising paradigm, the real-world social connection is used to communicate between the devices in the future, which could be called software-defined pedestrian network.

2.3.6. *Software-defined infrastructure's benefit recap*

Taking into account all the previous discussions, renovated SDI terminology possessed several benefits:

– The context boundaries between existing and uninvestigated software-defined inspired paradigms are removed.

– SDI's context is no longer limited in its ancient position but expanded to its latest required state (considering both physical and virtual perspectives, hardware deployment aspect).

Besides the above impacts on the literature, the novel SDI also promises to enhance the management aspect of the cutting-edge network infrastructure. Since any vehicle or network device equipped with SDI capability could participate and be managed by an SDI controller, a coping mechanism with heterogeneous hardware and software is presented. This feature allows open and unified collaborations and management between any network elements within the SDI ecosystem. Inspired by such advancement and 5G's service-oriented context, we concrete our paradigm by studying its effectiveness in various nonideal eHealth transport scenarios. Hence, healthcare communication demands a certain standard to satisfy the vital data transmission requirements. The proposed scenarios are categorized into two classes based on the vehicle's network point of view: individual perspective and supervision perspective. Before initiating our defined scenarios, we present a functional formalization model to control the prospects of SDI-inspired interventions.

2.4. Unified functional model formalization

As previously described, the SDI paradigm primarily offers the unification of networks with fully automated and adaptively self-manageable interventions in both software and hardware aspects. In this section, we formalize a unified functional model able to capture the versatility of the SDI hardware blocks and their collaborative prospects to improve the network performance.

2.4.1. *Description of the model*

Our proposed paradigm is based on a functional model with two main stages, as illustrated in Figure 2.5. The two-stage process (Mellouk et al. 2013) aims to optimize pseudo-static and dynamic network-related constraints to effectively deploy any potential SDI-empowered interaction to improve overall system performance.

In the first step, we perform a ranking of available strategies based on pseudo-static criteria. From all of the available solutions between a source-destination pair of considered strategies, we choose a subset of candidate strategies using a cost function. This function incorporates not only various static criteria (the use of surrounding third-party network equipments or the deployment of a dedicated network equipment such as UAV in order to meet a need or to monopolize all available resources, etc.) but also the application's needs and the environmental context. To allow adaptation of the strategy adopted in the first stage, QoS feedback is returned on a regular basis.

Then, in the second step, traffic is dynamically managed by the different infrastructure devices based on the strategy adopted in the first step. The weight assigned to each strategy depends on the evaluation of dynamic criteria such as packet loss, bandwidth and end-to-end delay. This evaluation requires defining a real-time cost function based on the considered dynamic criteria. For this function, we chose to adopt a reinforcement learning-based approach.

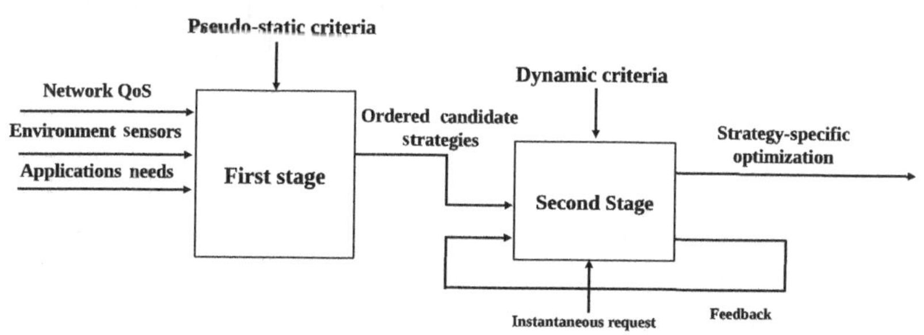

Figure 2.5. *Unified functional model of each SDI component*

2.4.2. *Structure of the model*

The model is built upon different modules, each of which optimizes a local cost function. To achieve this, we propose a unified formal model adapted from the framework proposed by

Bottou (2003) to determine the convergence of our approach. The foundation initially goes back to the works of Tsypkin (1973), which have been applied in various networking areas, in particular in routing solutions (Mellouk et al. 2013).

We estimate the convergence of our approach by observing the solutions of different stochastic equations converging to optimal values represented by $c = c^*$, where c represents the current weight parameters vector and c^* is the optimal state vector. x is denoted as the probability variable for the network-improving approach deployed by the controller. Hence, its probability density function is represented as $p(x)$, and $Q(x,c)$ is a random function for each realization of x.

Let us assume for a moment that $p(x)$ is known and that the functional of our system learning goal $J(c)$ can be written in explicit form:

$$J(c) = \int_X Q(x,c)p(x)dx \qquad [2.1]$$

Then combined with the assumption that $Q(x,c)$ is continuous with respect to c, the required condition for the functional extremum could be presented in the form of an equation:

$$\nabla J(c) = 0 \qquad [2.2]$$

In our approach, continuous learning entails modifying the network model's parameters while utilizing an adaptive update mechanism such as:

$$\Delta c[n-1] = -\Gamma(n)\nabla J(c[n-1])) \qquad [2.3]$$

This dynamic equation, which can be also expressed as a differential equation, serves as the model for obtaining the optimal vector $c = c^*$ as:

$$\frac{dc(t)}{dt} = -\Gamma(t)\nabla J(c(t)) \qquad [2.4]$$

The matrix Γ must be chosen in such a way that ensures the convergence of $c[n]$ or $c(t)$ to the optimal value c^*. Equation [2.3] encompasses numerous iterative formulas used in numerical analysis, but we will not expose them in detail here (Bottou 2003).

If there is insufficient prior information, that is, the probability density function $p(x)$ is unknown and cannot be estimated in advance, it is not possible to explicitly specify the condition in [2.2].

$$\nabla J(c) = M_x\{ \nabla_c Q(x,c) \} = 0 \quad [2.5]$$

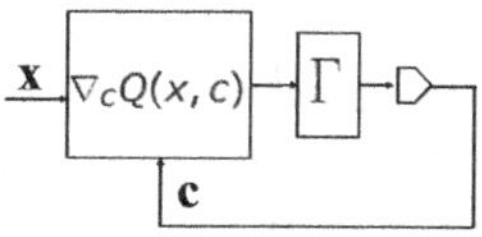

Figure 2.6. *Block diagram of the model structure*

In such cases, the optimal vector $c = c^*$ is determined progressively using learning algorithms. Learning algorithms are employed to generate an estimate of the vector $c[n]$ or $c(r)$ that gradually converges over time to the optimal vector c^*. These algorithms utilize observations of x, c and $\nabla_x Q(x,c)$ to refine the estimate. This is described in the block diagram of Figure 2.6, which describes a pseudo-autonomous system. The functional transformer $\nabla_c Q(x,c)$ has two inputs, one of which carries the current information and the other the pseudo-static criteria described earlier. In this case, the dynamic equation becomes:

$$\frac{dc(t)}{dt} = -\Gamma(t)\nabla_c Q(x(t), c(t)) \quad [2.6]$$

where

$$x[t] = x(nT) \quad when \quad nT \leq t < (n+1)T \quad [2.7]$$

is a stepwise function formed by discrete sample $x[n] = x[nT]$.

The success of the model can be guaranteed by achieving convergence, which refers to the convergence in the mean-square sense and/or with a probability equal to one. This convergence is reached when the function $c(t)$ produced by the model satisfies the following condition as demonstrated by Tsypkin (1973):

$$P\left\{\lim_{x\to\infty} [J(c(t)) - extr\, J(c)] = 0\right\} = 1 \qquad [2.8]$$

As explained in the previous discussion, our final step is studying the SDI effectiveness in the nonideal environment with strict network requirements. Consider the thesis context, we select emergency healthcare transport use case: ambulance transmission. More specifically, the ambulance must operate in an uncertain environment but required to send the data with utmost integrity and satisfy communication requirements (Qadri et al. 2020). Hence, we present a mapping prototype where the ambulance is the Q-learning agent to comprehend the connection between the formulated functional model and our use case. Inspired by the Q-learning approach, the ambulance will gradually learn a Q-table, a weighted matrix constructed by its feasible actions and current states, to navigate itself to the optimized approach. Since reliable ambulance transmission is the core concept in this case, SDI's interventions would be oriented to ensure this objective. Hence, the ambulance action space in the Q-table could be the set of precomputed SDI options (existing network slices, existing hardware or formulated route) or exploring new choices (ask for SDI controller supportive interventions). The current ambulance's network characteristics (bandwidth, delay, SNR, current hops) are candidates for formulating the agent state space. However, since each metric's space is continuous, we propose to discrete the formulated state space from a function that takes the metrics above as parameters. With the proposed Q-table structure, the ambulance will gradually discover the reward by measuring its current Quality of Service and updating the Q-table using the mentioned formalization approach. Once the overall QoS in the table is no longer significant, the ambulance succeeded in learning the weighted matrix and found its optimal approach. Table 2.1 and 2.2, and equation [2.9] illustrates the proposed reference example from the SDI formulated Q-learning model.

$$s_i = F_n(Q(B, D, N, H)) \quad [2.9]$$

where F is a predefined discrete function, B is average bandwidth, D is average delay, N is average SNR and H is the hops number.

RL system	SDI elements	Notation
- Environment	- Network environment	-
-Agent	- Ambulance	-
-State	- Discrete mapping of network characteristics category	s_i
-Action	- Routing decisions or demands SDI-intervention deployment	$a(s_i)$
- Policy	- Local routing decision	π
- Value function	-QoS estimation	$Q(B, D, N, H)$

Table 2.1. *Correspondence between a RL system and SDI's ambulance*

		Agent action		
		SDI-intervention 1	SDI-intervention j	Exploring (demand coverage or new interventions)
Agent state	State 1	Measured QoS Q_{11}	Q_{1j}	Q_{1m}
	State i	Measuring QoS Q_{i1}	Q_{ij}	Q_{im}
	State n	Q_{n1}	Q_{nj}	Q_{nm}

Table 2.2. *SDI Ambulance's Q-table structure*

Furthermore, we take advantage of SDI controller overview to aggregate the learned knowledge from each SDI-enabled device to formulate the ecosystem's unified functional model for SDI's ecosystem-level optimization. After finalizing the model aggregation, the SDI controller will forward the updated model to each SDI-enabled device that aims to improve its existing unified functional model. Figure 2.7 illustrated such internal SDI interactions to formulate the ecosystem's unified functional model.

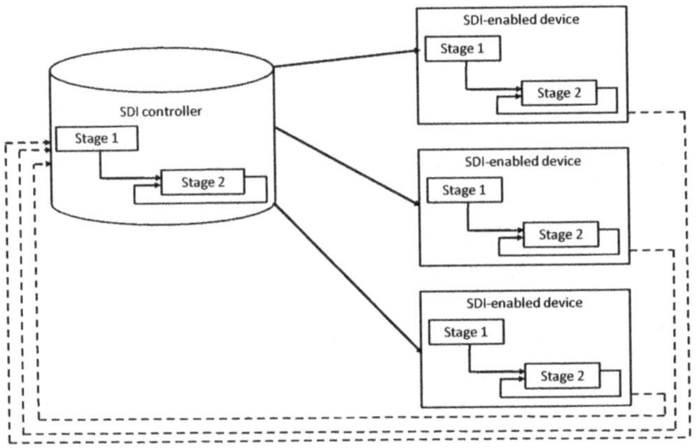

Figure 2.7. *SDI ecosystem unified functional model*

After presenting the prototype mapping, we present four intervention conjunctions as SDI-empowered strategies toward overall network performance for the evaluation purpose. The following section describes our experimental environment and proposed scenarios for such procedures.

2.5. Description of the experiments

Inspired by conventional previous SDI interactions and medical emergency conditions which require elastic infrastructure, an experimental environment in critical circumstances is formulated to validate our proposal. Motivated by the scenarios proposed in Campolo et al. (2018) and the chapter's context, we consider an ambulance carrying its patients to the hospital. The ambulance maintains its livestream healthcare vital data feed to the nearest edge-computing server to have the best medical assistance, as shown in Figure 2.3.

Taking into account the individual network context, every vehicle, especially the ambulance, should be able to use its network view to enhance its network performance. Thanks to SDI's openness, the SDI technology awareness in each vehicle could allow opportunistic network enhancements within the area. Combining two previous factors, two first SDI improvement approaches are introduced: opportunistic coverage enhancement and recovery connection.

Along with the existence of an opportunistic vehicle, such criteria may not be guaranteed in reality. Hence, we propose to embed a dedicated UAV into the ambulance to enhance its SDI versatility. This dedicated hardware allows a flexible, independent advancement versus the opportunistic approach. However, working network infrastructure is still mandatory to ensure the network functionality. Thus, the third SDI enhancement approach still takes advantage of individual network points of view but is categorized as a semi-independent solution. This approach's name is self-assisted coverage enhancement and data transportation. Considering other cases, each vehicle's network point of view and solitary managing capabilities are insufficient to handle the circumstances to benefit the ambulance network's performance. A central supervisor's attendance and prioritized orchestration is obligatory to resolve the above problem. Therefore, we take advantage of SDI's controller supervision as the final solution in our last SDI improvement approach, "priority orchestration in high-density network".

Before returning to each approach's detailed introduction, the readers should notice some highlights of the framework used for simulating the experimental environment.

2.5.1. *Implementation details*

2.5.1.1. *Experimental framework*

Mininet Wifi (Fontes et al. 2015), the popular container-based network simulation tool with extensive mobility models, wireless network capability and software-defined functionality, is the most suitable environment for our various experiments.

However, Mininet Wifi's current implementation does not fully support our proposed idea. Since the SDI idea enables the capability to deploy network infrastructure and those deployed hardware's network and location are managed by the SDN controller, the network hardware is supposed to be mobile and SDN-manageable. The simplest approach is combining Openvswitch's SDN functionality and Mininet Wifi's moving vehicle equipped with wireless interfaces. However, such mixture in the Mininet WiFi's latest implementation is not ready to use. Thus, we proceed to embed Openvswitch's OpenFlow capability on the mobile vehicles that enable their network stack to be managed by an SDN controller in our framework. In conclusion, an essential Mininet WiFi extension is implemented to initially enable SDI functionality to support our experimental requirement. The upgraded Mininet Wifi with SDI-enabled capability can be obtained by contacting the authors.

After the simulation upgrades, we chose the Ryu's (Asadollahi et al. 2018) framework to implement software-defined controller functionalities in our initial solution implementation.

2.5.1.2. *Experimental scenario inspiration*

As inspired by the experiment architecture in Campolo et al. (2018), the chapter's context and illustrated earlier in Figure 2.3, the hardware elements required to compose our scenarios are as follows. High-performance cloud architecture is used to manage the whole network's activities. Static core SDN switches are intermediate connectors that provide high-performance network functionalities between the backbone network and end-user access points. However, multi-access edge-computing servers are located near the end-user rather than the cloud, various types of modern equipment (UAVs, vehicles, etc.) provide coverage and formulate their communication infrastructure. Both ancient and cutting-edge base stations (with smaller ranges) will ensure broad and flexible signal coverage, leading to the nearest edge computing server and efficient SDI enhancements.

Taking advantage of communication capabilities offered by the outlined architecture, the ambulance maintains its patient's eHealth data stream to the edge server while on its course to the hospital. Each scenario will characterize variant interventions between the ambulance and the infrastructure to improve the circumstances.

Each approach's description will be provided regarding their names and illustrated in dedicated figures. The standard structure of each description will be composed of four parts on the reader's behalf: the communication infrastructure's elements, scenario overview, triggered events and SDI-empowered interventions. This structure allows the readers to quickly identify and locate the discussion result based on their interests. The essential timestamps are also marked to highlight differences between the vital interaction empowered by SDI and non-SDI approaches.

Further analysis in section 2.6 will respect the same structure by discussing each highlighted event and its affections on obtained results. Table 2.3 summarizes the experimental method summary for each approach.

Approach	1. Opportunistic coverage enhancement	2. Connection Recovery	3. Self-assisted Coverage Deployment And Data Transportation	4. Priority Orchestration in High-density network
Communication infrastructure's elements	A1 and A3 are cutting-edge base stations. B1 is a more ancient base station with larger coverage and poorer coverage quality.	B1 and B2 stations.	Station B1 provides signal coverage as access point infrastructure. An SDI-enabled UAV is embedded in the ambulance.	B1 and B2 are ancient base stations without SDI capability. A2 is modern and SDI-enabled, thus employing the traffic-orchestrated capability.
Scenario overview	The ambulance passes through the network infrastructure's covered area. A vehicle located near the position of station B2 in both sub-scenarios but only SDI-enabled in the second one.	While in station B1's coverage and reaching to B2's, B1's malfunction occurs and the ambulance lost its communication capability.	The ambulance lost its network infrastructure suddenly, and no backup nor an opportunistic device existed in the area. A SDI- enabled high-speed UAV embeds in the ambulance is deployed, carries a part of crucial data and improves coverage enhancement.	A significant amount of vehicles also enter the area and consume the network resources that cause the decreasing in the ambulance's network performance.

Approach	1. Opportunistic coverage enhancement	2. Connection Recovery	3. Self-assisted Coverage Deployment And Data Transportation	4. Priority Orchestration in High-density network
Triggered events.	#1: The ambulance enters station B2's area. #2: The ambulance exits station B2's area.	#1: B1's malfunction moment #2: Temporal SDI formulation. #3: Ambulance reaches station B2's coverage.	#1: UAV reaches the infrastructure and delivers urgent data to the edge server. #2: The ambulance enters UAV's enhanced coverage. #3: The car enters B1's coverage area.	#1: The moment other vehicles also enter the B1's covered area while the ambulance is still in it. #2: The ambulance exits B1's coverage and changes its next base station.

Approach	1. Opportunistic coverage enhancement	2. Connection Recovery	3. Self-assisted Coverage Deployment And Data Transportation	4. Priority Orchestration in High-density network
SDI-empowered interventions.	Right after the ambulance enters the vehicle's covered area, the ambulance acknowledges the existence of a SDI vehicle. SDI capability allows two vehicles to construct a temporary software-defined infrastructure and take advantage of communication resources to boost in the ambulance's network performance.	The ambulance takes advantage of existing vehicle in nonsignal area to maintain its communication capability.	The ambulance is aware of the non-infrastructure situation. It deploys the embedded UAV to carry the specified vital eHealth data. The UAV reaches the nearest base station, then prioritize to forward the critical data and activate the scanning mode to wait for the ambulance. Once the ambulance is in the UAV's extended coverage, the UAV senses it and then activates the forwarding mode.	The UAV reaches the nearest base station, then prioritize to forward the critical data and activate the scanning mode to wait for the ambulance. Once the ambulance is in the UAV's extended coverage, the UAV senses it and then activates the forwarding mode.

Table 2.3. *Experimental approach context summary*

2.5.2. *Approach 1: opportunistic coverage enhancement*

The preliminary executed scenario is opportunistic coverage enhancement, inspired by the network slicing collaboration mechanism (Campolo et al. 2018). Figure 2.8 illustrates the scenario architecture. Three network metrics (throughput, delay and packet loss) will be measured to illustrate SDI enhancement. The obtained results and subsequent analysis will be described in section 2.6.

2.5.2.1. *Communication infrastructure's elements*

The communication signal for the ambulance will be provided by three base stations: A1, B2 and A3.

While stations A1 and A3 are cutting-edge base stations equipped with the most recent communication technology, hence having the best communication performance, Station B1 is a more ancient base station. Specifically, B1's signal coverage area might be larger, but the coverage quality is more bounded than A1 and A3.

2.5.2.2. *Scenario overview*

The ambulance passes through the network infrastructure's covered area. The ambulance conducts its information process to the edge server and initiates handover each time it switches to the base station. A vehicle located near the position of station B2 in both sub-scenarios but only SDI-enabled in the second one.

2.5.2.3. *Triggered events*

Event markers indicate moments when the ambulance enters (t = 45 s at event #1) and exits (t = 60 s at event #2) station B2's area.

2.5.2.4. *SDI-empowered interventions*

The first SDI-empowered interactions occur right after the ambulance enters the vehicle's covered area. Since the ambulance acknowledges the existence of a SDI vehicle, SDI capability allows two vehicles to construct a temporary software-defined infrastructure and take advantage of communication resources. The provisional advancement allows a boost in the physical layer of the ambulance, which influences its network performance temporarily until the ambulance is out of range from the vehicle.

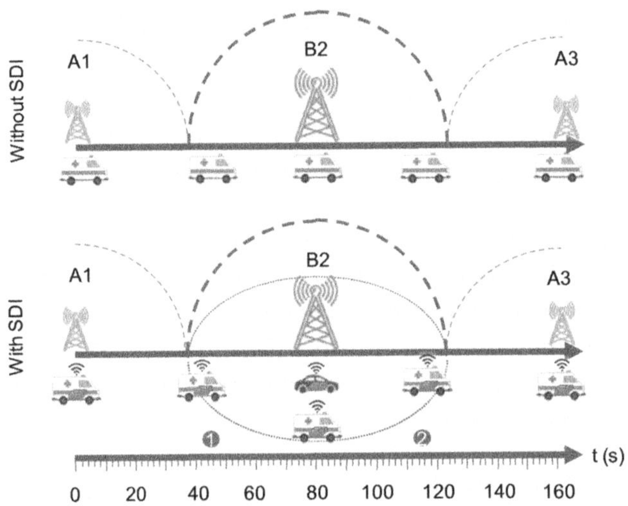

Figure 2.8. *Scenario 1: opportunistic coverage enhancement. For a color version of this figure, see www.iste.co.uk/chu/softwaredefined.zip*

2.5.3. *Approach 2: connection recovery*

In the first circumstance, poor signal or ancient technologies may degrade the network performance. However, the infrastructure still functions properly and does not introduce any service disruption. In critical situations correlated with the work's context, a particular incident, such as a power accident, could happen and cause some of the healthcare communication infrastructure to shut down. In this case, if the affected base station is the only one in the area, the ambulance must have a network infrastructure to rely on. The eHealth data feed would not be able to survive until the ambulance reaches a new signal-covered area. However, this limitation could be eased thanks to SDI. The second architecture, which stimulates the scenario in Park and Yoo (2017), is shown in Figure 2.9.

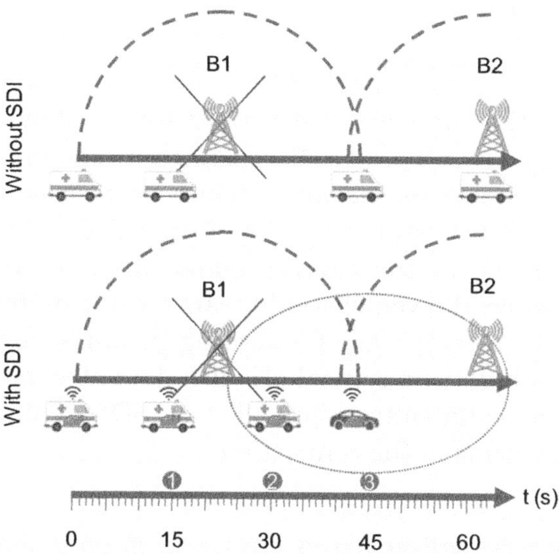

Figure 2.9. *Scenario 2: connection recovery. For a color version of this figure, see www.iste.co.uk/chu/softwaredefined.zip*

2.5.3.1. *Communication infrastructure's elements*

Stations B1 and B2 will provide the signal coverage.

2.5.3.2. *Scenario overview*

The ambulance is passing by B1's area and is expected to reach B2 afterwards, then suddenly, a power incident happens at B1, causing B1's malfunction later.

The ambulance is required to wait until it reaches station B2 to restore communication.

2.5.3.3. *Triggered events*

The power incident happens at t = 15 s and is marked by the event #1's marker. Event #2 indicates when the ambulance acknowledges an opportunistic SDI-enabled vehicle. The final marker (event #3) occurs when the ambulance reaches station B2.

2.5.3.4. *SDI-empowered interventions*

Right after the network infrastructure is down, the ambulance takes advantage of SDI's openness to communicate with existing vehicles in the area, which is also equipped with SDI technology. Fortunately, one of the existing vehicles in the area may be in the covered signal area by another base station. Hence, the vehicle could transform itself from an end-user node to a temporary forwarding network device to maintain the ambulance's critical eHealth data stream to the edge server. This collaboration provided by SDI significantly reduces the downtime of the connection.

2.5.4. *Approach 3: self-assisted coverage deployment and data transportation*

The two above scenarios share a common characteristic: an SDI-supported vehicle plays as a temporary forwarding node to improve or even establish indispensable network coverage. However, such criteria may not always be valid in practical conditions, hence the third scenario is conducted. Figure 2.10 demonstrates the experimental architecture in this scenario, which is motivated by a delay-tolerant network in Dias et al. (2017).

2.5.4.1. *Communication infrastructure's elements*

Station B1 provides signal coverage as access point infrastructure. An SDI-enabled UAV is embedded in the ambulance.

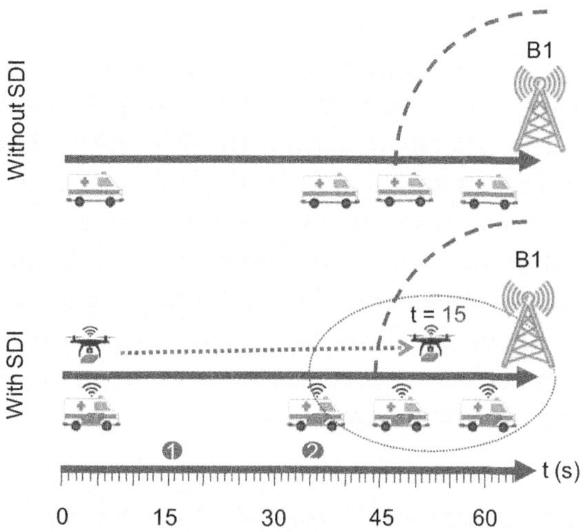

Figure 2.10. *Scenario 3: self-assisted coverage deployment and data transportation. For a color version of this figure, see www.iste.co.uk/chu/softwaredefined.zip*

2.5.4.2. *Scenario overview*

The ambulance loses its network infrastructure suddenly, and no backup nor an opportunistic device existed in the area. The ambulance is required to recover the connection, in essence. However, in the first sub-scenario, the ambulance has no solution to exploit. In the second sub-scenario, the ambulance is embedded with an SDI-enabled high-speed UAV that can carry a part of crucial data and enhance network coverage.

2.5.4.3. *Triggered events*

The event #1 (t = 15 s) marks when the UAV reaches the infrastructure and delivers urgent data to the edge server. The event #2 occurs at t = 35 s when the ambulance enters the UAV's enhanced coverage. Once the car enters B1's coverage area (t = 45 s), event #3 happens.

2.5.4.4. *SDI-empowered intervention*

After the signal detection phase, the ambulance is aware of the non-infrastructure situation. It deploys the embedded UAV to carry the specified vital eHealth data. The UAV reaches the nearest base station by t = 15 s, then prioritize to forward the critical data and activate the scanning mode to wait for the ambulance. Once the ambulance is in the UAV's extended coverage, it senses the ambulance and activates the forwarding mode by t = 35 s. This action allows the ambulance's connectivity recovery, thus significantly reducing the eHealth data stream's downtime.

2.5.5. *Approach 4: priority orchestration in high-density network*

The last but not least important scenario is priority orchestration in high-density network, inspired by the hybrid approach in Toufga et al. (2018). The network infrastructure functionalities might be sustained across its entire operational cycle.

However, a burst in the density of devices may result in higher network resource consumption and introduce degradation to the quality of service (QoS) for the rest of the infrastructure. In such a case, our last scenario indicates the degradation in network performance in the ambulance and the other vehicle while sharing the same base station without and with SDI capabilities. Experimental architecture is shown in Figure 2.11.

2.5.5.1. *Communication infrastructure's elements*

In this scenario, three base stations, B1, B2 and A2, will be used to provide coverage signal. While B1 and B2 are ancient base stations without SDI capability, A2 is modern and SDI-enabled, thus employing the traffic-orchestrated capability.

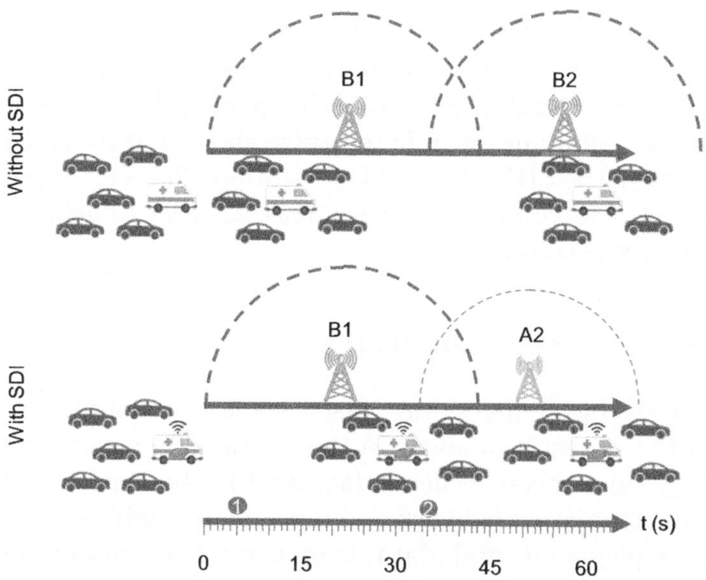

Figure 2.11. *Scenario 4: priority orchestration in high-density network. For a color version of this figure, see www.iste.co.uk/chu/softwaredefined.zip*

2.5.5.2. *Scenario overview*

While the ambulance is moving through the covered area provided by stations B1 and B2, a significant amount of vehicles also enter the area and consume the network resources. Since a high network demand occurs and no priority mechanism is implemented in ancient base stations, the ambulance's network performance is decreased based on the number of competing vehicles.

2.5.5.3. *Triggered events*

The event marker #1 indicates the moment other vehicles also enter the B1's covered area while the ambulance is in it. Event #2 happens when the ambulance exits B1's coverage and changes its next base station.

2.5.5.4. *SDI-empowered intervention*

In this scenario, no intervention is initiated from the SDI vehicles. However, such intervention happens at base station A2. Once the ambulance is able to utilize the signal provided by A2 (t = 35 s), the SDI-empower base station allows it to take advantage of reserved network resources in order to achieve better network performance.

2.6. Improvement evaluation approaches

After multiple experiment discussions, we shift our attention to the result analysis in this section. To ensure the readability, our analysis order is identical to the experiment order: opportunistic network enhancement, self-assisted coverage deployment and data transportation, connection recovery and priority orchestration in high-density network.

Our obtained results are plotted from Figures 2.12–2.15(b). We improve the data's readability by applying the average window technique to smooth the data with a window size equal to 5.

A clear pattern that could be observed in all of the figures is the presence of distinctive data groups where a significant change in network performance became manifested. This phenomenon indicates the association between the triggered events per each scenario as well as their effect on the actually collected network metrics. Each scenario analysis below will take into account such transitions to highlight SDI's advancement.

2.6.1. *Opportunistic coverage enhancement*

Figure 2.12 depicts the ambulance's network performance in the first scenario in terms of three network metrics: bandwidth, delay and packet loss.

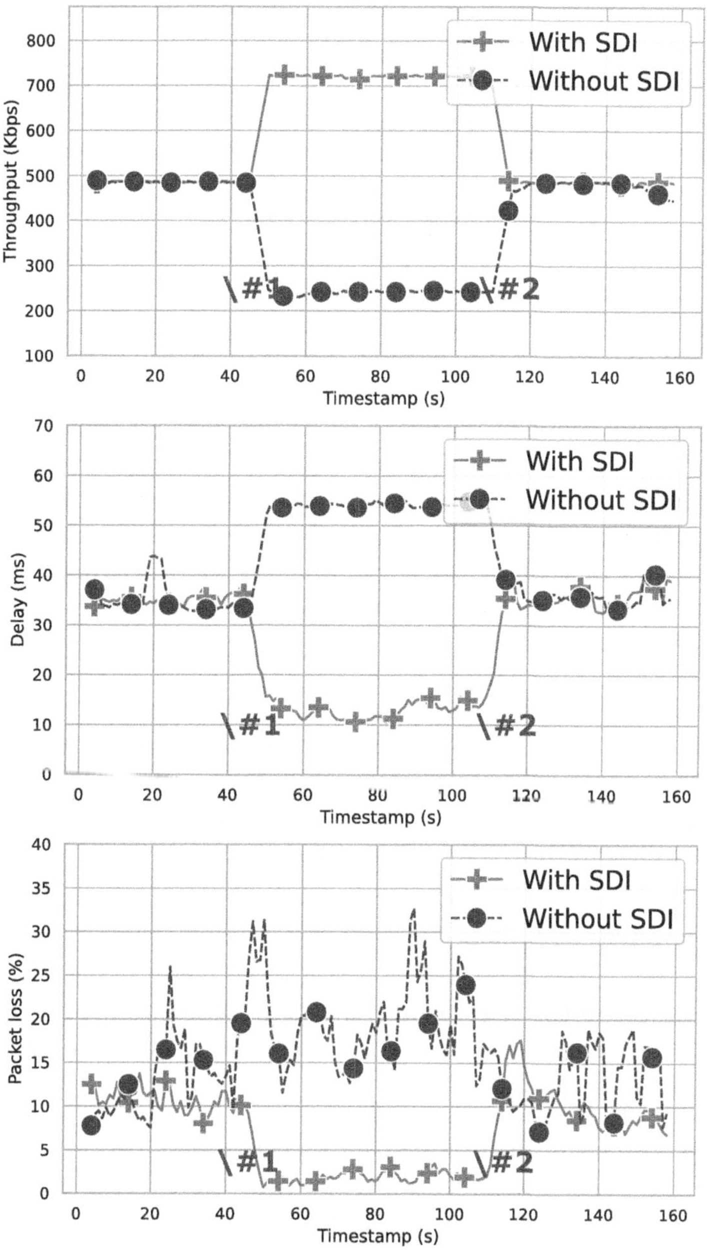

Figure 2.12. *Network metrics. For a color version of this figure, see www.iste.co.uk/chu/softwaredefined.zip*

The evident trend in the three subfigures is the falling (without SDI solution) and the rising (with SDI solution) in terms of the ambulance's network performance in the figure's midsection. The data shifts happen at t = 45 s and t = 115 s regarding the ambulance's signal influence by base station B2 and SDI-enabled vehicle once it enters the path's central.

The ambulance's network performance declines once it is in B1's coverage because station B1 provides a lower quality of network coverage compared to more advanced base stations A1 and A2.

In the equivalent situation, because of SDI intervention, the ambulance can take advantage of an opportunistic SDI-enabled vehicle to enhance its coverage under the ancient infrastructure. This leads to the network metrics' improvement increasing twice versus the period when the ambulance was being served only by the cutting-edge base stations A1 and A3.

Considering the mentioned transitions between the three collected metrics (throughput, delay and packet loss), which share the same characteristics in other scenarios, the throughput remains the only metric to be plotted in further analysis to highlight SDI's enhancement.

2.6.2. *Connection recovery*

Equivalent to the first scenario, in the connection recovery scenario, the throughput from the ambulance is measured and plotted in Figure 2.13.

Three apparent data transitions correspond respectively with three following events: B1's shutdown (t = 15 s), connection recovery provided by SDI-enabled vehicle (t = 30 s) and connection recovery provided by traditional infrastructure (t = 45 s). The data transitions exhibit the main difference between the two circumstances, where the SDI solution possesses a lower connection recovery time than its counterpart.

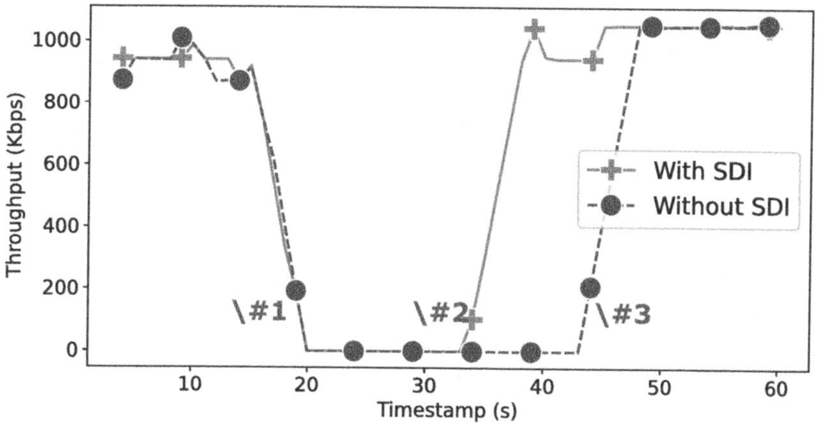

Figure 2.13. *Network bandwidth in connection recovery experiment. For a color version of this figure, see www.iste.co.uk/chu/softwaredefined.zip*

Regarding the experimental process, the throughput is stable from t = 0 s to t = 15 s since the network infrastructure functions properly. Then it drops to 0 after the power incident happens, causing the network to become inaccessible.

Without SDI, the ambulance could not recover the connection to the edge server until it reached the new base station by t = 45 s.

In the second scenario, with the SDI's advantage, the ambulance senses the existence of SDI-enabled vehicles at t = 30 s, hence establishing the collaboration to restore communication capability. This mechanism reduces significant downtime, allowing minimum critical data streaming disruption.

2.6.3. *Self-assisted coverage deployment and data transportation*

Unlike the two previous scenarios where the SDI's improvement originates from opportunistic vehicles, an SDI-enabled UAV is utilized to transport prioritized data, deliver it ahead of time and improve the ambulance's signal reachability in this scenario. Thus, the data received by the edge server is from the ambulance and the UAV's delivery. Thus, using solely the throughput from the ambulance perspective does not depict the experiment overview. For this reason, we keep measuring the throughput of the eHealth data stream. However, the viewpoint for capturing the SDI intervention process is switched from the ambulance to the edge server. The obtained result is illustrated in Figure 2.14.

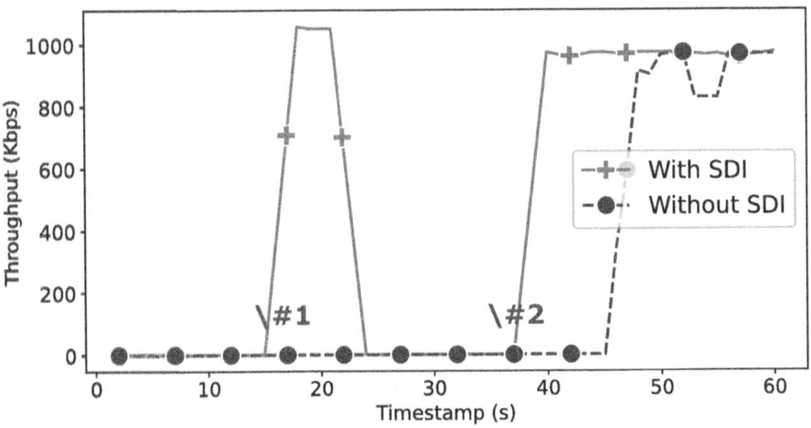

Figure 2.14. *Network bandwidth in self-assisted coverage enhancement. For a color version of this figure, see www.iste.co.uk/chu/softwaredefined.zip*

The significant data drifts could be identified as correlated with triggered events in this scenario description: the edge server receives a proportion of data around (from t = 15 s to t = 25 s), and the precedence of data arrival in SDI-enabled sub-scenario (t = 35 s) versus non-SDI sub scenario (t = 45 s).

Initially (t = 0 s), the throughput equals zero, thus the infrastructure was downed, and the ambulance could not connect to the edge server.

Without SDI intervention, the edge server typically receives no communication until the ambulance enters the new base station's coverage area (t = 45 s).

Meanwhile, in the second scenario, the server received a vast amount of data significantly earlier (t = 15 s). These data come from the ambulance's embedded UAV, which carried the critical information and transmitted it as soon as it reached the working infrastructure. This UAV also enables significant downtime reduction for the ambulance when recovering its network coverage at t = 35 s.

The improvement of embedding an SDI-enabled UAV in this section could be summarized by prioritizing a partial vital data delivery and temporarily increasing the ambulance's reachability. However, from a more abstract perspective, the SDI-enabled UAV could be considered a dedicated and independent ambulance solution compared to previous approaches supported by third-party vehicles.

2.6.4. *Priority orchestration in high-density network*

The common concept between precedent scenarios is the source of SDI interventions from network nodes and their transformations to become the network infrastructure to cope with the communication obstacles. However, using one node's restricted view may not be adequate to resolve the holistic network problem that requires a global network view.

As aforementioned in the experiment description, the experiment's purpose is to reconstruct network behavior when a burst in network demand is caused by a high-density network that negatively affects the ecosystem's overall network performance. Hence, the collected metric in this scenario is the ambulance's throughput and the other vehicles. Because of the involvement of numerous third-party vehicles, two distinguished figures are used for comprehensive illustration. While Figure 2.15(a) presents the bandwidth of the ambulance and the remainders' without SDI's influence, Figure 2.15(b) is for the scenario that includes the ambulance's network enhancement impacted by SDI interactions.

Similarly to the previous experiment, two particular data transitions that deserve the readers' attention are the dramatic decrease in the ambulance's throughput at t = 5 s, and the fluctuation at t = 35 s in both subfigures.

In the beginning (t = 0 s), the ambulance's throughput in both figures is initiated at the highest level and the others is zero. This illustration is because for the first 5 s, the third-party vehicles have not yet entered the coverage area provided by the first base station. The ambulance's throughput decreases significantly once the other vehicles consume the B1's resources. The metric keeps declining until the ambulance is served by new base stations B2 and A2. However, under B2's management, only a short rise in the ambulance's throughput is witnessed since the other vehicles still need to be served by B2. As soon as the others enter B2's area, the rise disappears, and the ambulance continues to experience network resource shortage, which results in low measured throughput. It is necessary to have a solution to guarantee the ambulance's throughput to keep the eHealth service intact.

Mandating sufficient network resources for high-priority transmission requires the system overview to satisfy the requisite network management. SDI-enabled base station A2 will handle the situation more effectively in the same condition with its supervision. The centralized controller performs the function of ensuring priority in the cellular operational environment.

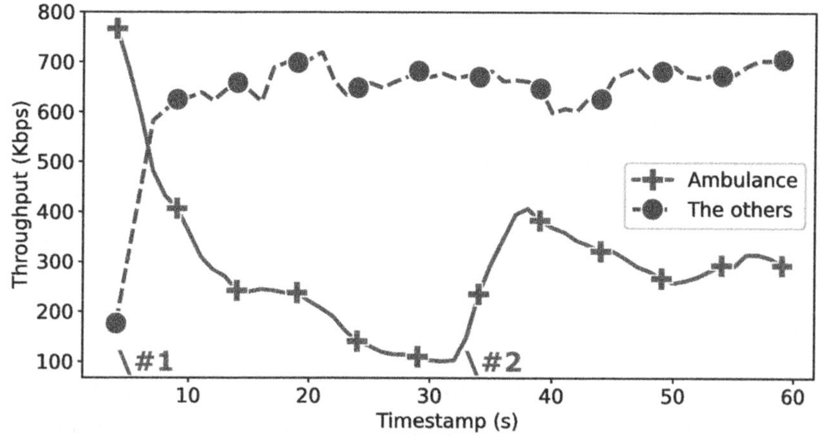

(a) Network throughput in priority orchestration in high-density network without SDI

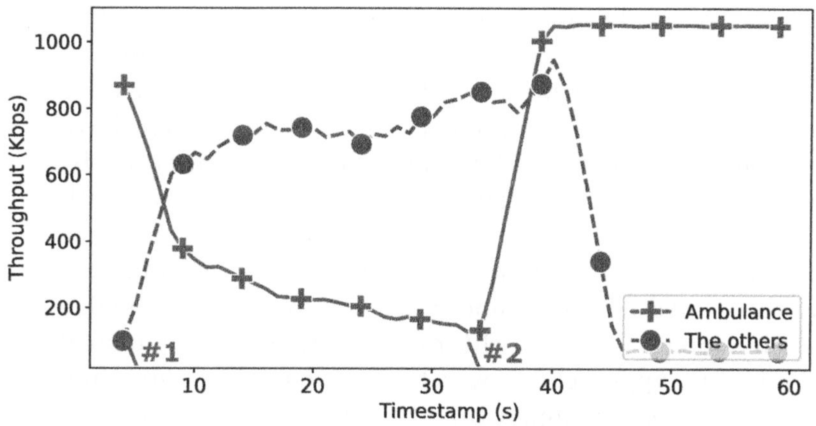

(b) Network throughput in priority orchestration in high-density network with SDI

Figure 2.15. *Network throughput in priority orchestration in high-density network. For a color version of this figure, see www.iste.co.uk/chu/softwaredefined.zip*

When the ambulance reaches the SDI-enabled base station (t = 35 s), it receives the highest priority to access the network resource to maintain its QoS requirements. Hence, the throughput returns to the highest. Meanwhile, the third-party vehicles' performance is limited within a threshold.

Thanks to SDI-enabled interventions, the raised concern about individual network nodes' views, which may not be sufficient to be an effective solution for the holistic problem, has been resolved effectively.

2.7. Conclusion

In summary, this work addresses the existing pitfall of the community research directions and the industry vision in an SDN paradigm in the healthcare domain, which is the deficiency of the modern network infrastructure's evolution. We tackle this issue by extending "software-defined infrastructure", releasing it from outdated understanding in the literature and granting it the most recent perspective and future real-world applications. Crucially, we have shared our vision of the following technology generation by elaborating the cutting-edge 5G network perspective with the SDI terminology extension. A unified, flexible, dynamic and automatable network infrastructure is formulated from the proposed idea, encompassing most cutting-edge research topics in the literature. Four distinguished experimental approaches, which consider the individual network perspective (opportunistic network enhancement, self-assisted coverage deployment and data transportation, and connection recovery) and supervision perspective (priority orchestration in high-density network), are also designed to facilitate promising enhancements by the internal SDI ecosystem perspective that aims to support the domain's most challenging scenario: vehicular healthcare emergency communications.

Besides, while validating our proposed idea, we also extend the cutting-edge simulation tool, Mininet WiFi, and include a cutting-edge SDN controller to initiate a framework for further investigations from the community in this research direction. This work also included our complex and extensive validation results to confirm the benefits of our suggested proposal in terms of network metrics (bandwidth, delay and packet loss) in our various designed scenarios. Such benefits promise significant improvements in network aspects, not to mention the proposal's broad range of applications that require strict communication conditions similar to healthcare emergencies (autonomous driving, virtual reality, etc.). Nevertheless, the presented work is the beginning effort to open the door for a novel vision of network infrastructure especially in H-IoT but could be extend to general IoT context or in other domains.

After the first contribution, the following chapter will present another solution focusing on healthcare emergencies. The proposed solution takes advantage of the SDI's network overview and accessibility of two network endpoints to provide a seamless network experience for the ambulance in a dynamic and complex environment.

Chapter 3

SMART Connection Migration

Stateful application relocation

Barbarulo et al. (2022)

This chapter presents a novel solution for a seamless and transparent healthcare transport network experience in a dynamic and complex environment.

3.1. Introduction

Nowadays, advancing cutting-edge network technologies are one of the most critical factors that enable the success of modern information system applications (social media, media streaming services, interactive online workspaces, etc.). These technologies, such as 5G (Benelmir et al. 2022), cloud computing (Mushtaq et al. 2016), edge–fog computing (Bitam et al. 2018) and software-defined networking (SDN) (Bannour et al. 2022), not only enable scalability (with billions of worldwide users) but also enhance the end-user's accessibility to diverse software platforms. In addition to the Internet infrastructure's evolution, network protocols need to evolve to meet end-user expectations, which have become increasingly demanding, especially in mobile environments. However, old-fashioned, albeit successful, transport protocols such as TCP and UDP cannot support such requirements in their initial design. The original protocol neglected the possibility of users continuously experiencing network disruptions in a highly dynamic environment, thus causing user frustrations. The importance of re-establishing a connection each time an end device changes its network environment should not be underestimated. QUIC (Iyengar and Thomson 2021) and its multi-path variant MP-QUIC (Chu et al. 2022, 2023) were created as successor transport protocols with several innovative highlights. One of these is the novel *connection-id* parameter, which allows QUIC-based servers and clients to maintain existing connections even when the middle network infrastructure suffers frequent network changes.

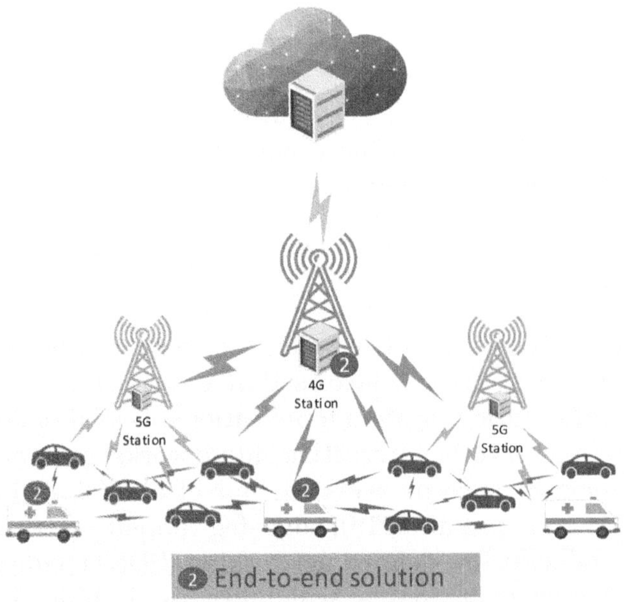

Figure 3.1. *Second contribution's prospect in the overall framework. For a color version of this figure, see www.iste.co.uk/chu/softwaredefined.zip*

In the context of this work, regarding the strict communication requirements of the H-IoT domain, this feature could play an essential role in ensuring the success of transmitting vital healthcare data over dynamic network infrastructure. An example of this case could be an emergency ambulance, which can maintain its vital data transmission to provide prehospital care for the patient while on its way to transport the patient to the hospital. Besides maintaining patient network accessibility, the feature could allow the proposed SDI paradigm to enhance the ambulance's network performance by migrating the current service to the optimal edge servers after activating the SDI's unified functional model. The presented prospect of this chapter's contribution is also illustrated in Figure 3.1.

While the enhanced capability made its name under the "connection migration" terminology for over a decade, its feasibility was not available natively until the emergence of QUIC-based protocols. Observing its historical path in Sultan et al. (2002), Tan et al. (2020) and Conforti et al. (2021), "connection migration" has recently become achievable (Conforti et al. 2021; Sisinni et al. 2023), thus enabling the practicability of the previous work series in Machen et al. (2017) and Ma et al. (2018). The approaches proposed in Conforti et al. (2021) and Sisinni et al. (2023) (Explicit and Pool of Address [PoA]) offer the client straightforward solutions to maintain its connection to the new server. The work investigated the server-side "connection migration". Meanwhile, the connection migration process depends on the existing server's narrow view (Explicit) or a set of pre-defined servers (PoA), which requires additional effort from the client side to execute the connection migration function. The simple approach proved the paradigm's feasibility but needed to cope with the complex system infrastructure's dynamicity regarding the context of this work. Despite the authors' overview in this research direction, their proposed solution still needs an update to fully exploit the paradigm's potential and make it a "smarter" solution. To be more precise, we are suggesting a "third-party connection migration" approach in this book. This update will allow the connection migration not only from the client or server side but also from third-party entities (e.g. SDN or SDI controller) while considering various decision criteria. Enriched by the information from the network's current state and the infrastructure's internal operation, which has always been dynamic and unpredictable (Van et al. 2018; Tchernykh et al. 2019), a third-party approach plays its role in guaranteeing a seamless and transparent network experience for on-demand clients, which converges to the final purpose of the paradigm.

Since we have pointed out the limitation in the state of the art (SOTA) and suggested the necessary enhancement, we also introduce our initial efforts to fill this research gap. Our contributions toward a third-party connection migration solution could be encapsulated as follows:

– design and evaluation of a third-party connection migration approach and proof-of-concept mechanism;

– evaluation in a simulated environment: impact of RTT and migration frequency on the connection migration time.

The rest of this chapter is structured as follows. We present the background, related material and the motivation that inspires our solution in section 3.2. Section 3.3 considers the details of our solution design, while the experimental setup, obtained results and in-depth discussions are presented in section 3.4. Finally, we present our study's conclusions in section 3.5.

3.2. Related work

First, this section presents our source inspiration and the "connection migration" terminology. We then discuss step by step the technological landscape that enables the paradigm's feasibility. After this, we point out the limitation of the SOTA, then the motivation that inspired our motivations for a "third-party connection migration", which later led to the smart solution for a connection migration.

3.2.1. *Virtual machine migration*

As per the study by Silvera et al. (2009), the need for transparently maintaining the software service to the end users has appeared in the primitive form of migrating the whole virtual machine (VM). Silvera et al. proposed a VM migration scheme to support migrating one VM into a new host location while keeping its previous IP (to continue serving existing clients). The proposed scheme requires two additional actors to execute the task: two subnet agents and a migration manager. Since that original VM must keep its original IP, no other information in the network layer could be used to identify the destination's location. Thus, using two subnet agents to handle the routing and direct the migration process is inevitable. So the drawback in the VM migration approach is clear: it is mandatory to keep the same IP address to maintain existing connections, even if the VM changes its host location.

3.2.2. *Container migration*

Thanks to the advances in container technology, a lightweight approach for managing multiple isolated services has been widely adopted in the community (Barbarulo et al. 2022a, b). Instead of requiring a middleware between the running code and the hardware like in the VM approach, container technology allows different services to take advantage of the bare metal performance while still being managed efficiently and separately within a shared hardware and a common operating system. A significant improvement in performance gain has also been reported in several studies (Sisinni et al. 2023; Manoj Kumar et al. 2022). Being lightweight, containers can be transferred across a network in a very short time, thus improving migration time. However, the lack of distinction between network-level and application-level connection in TCP's design becomes harder to cope with. Since only the container is migrated to the new destination, which is in a different location, the service's IP address will necessarily change. As a result, all TCP-based established connections must be terminated, then re-established. This behavior seems the opposite expectation in the service migration paradigm regarding the continuity characteristic's network experience. The following description will explain the community's effort toward a universal solution.

3.2.3. *TCP-based connection migration*

Looking at the philosophy of the TCP connection, to ensure the ability to identify a process in the complex network infrastructure, it will be determined using two pieces of information: first, the IP location of the endpoint, and second, the port number associated with that process. A combination of two endpoints represents a TCP connection. The identification information changes whenever the underlying system changes its current IP address or associated port, thus making the current endpoint obsolete. Since the original combination no longer exists, the current connection breaks and must be terminated (Yu et al. 2023). In TCP, the idea of maintaining the same logical connection between two endpoints only exists if both endpoints keep their IP and the port remains unchanged throughout the communication process.

Note that the work by Sultan et al. (2002) was one of the first works to try to support the connection migration capability in TCP. Sultan et al. proposed an extension for the well-known protocol. The mechanism is triggered from the client side, taking advantage of the client's perspective to migrate to a pre-defined server. Multiple handshake interactions between the client, the existing server and the new server are mandatory to ensure the mechanism's effectiveness. The approach neglected the original TCP's philosophy about the unbreakable bond between endpoint identification and the logical connection by changing the connection's destination IP address.

The TCP's connection concept has been serving its intended purposes throughout the Internet's history, where the network components used to evolve very slowly. However, this is no longer true nowadays. The platform fluctuations in the cloud and edge environments (Kumari and Kaur 2021), small cell network deployment in 5G and 6G (Chen et al. 2023), and especially the rising mobile traffic trend are three main contributors to the rapid evolutions of the infrastructure. Even if the same two processes are still trying to communicate, any changes in the identification information definitely break the TCP connection. Therefore, the general Internet connection concept requires an update to converge with computer network developments.

3.2.4. *QUIC-based connection migration*

Regarding the need to revise the Internet connection concept, Google introduced the QUIC protocol in 2012, the goal of which, among many others, was to tackle this issue. By adopting a novel parameter called "connection-id", QUIC separated the perceptions between the two endpoints' logical connection and the underlying OS-level connection. This approach instantly resolves the existing service migration lock by allowing both the source and the destination IP address to be changed without breaking the connection flow. Meanwhile, the mechanism also ensures the living container migration's feasibility while maintaining the established QUIC-based connections. Hence, the client-side connection migration mechanism is natively supported in the protocol. As mentioned in the introduction, service migration schemes proposed in Tan et al. (2020) and Conforti et al. (2021) are all achievable thanks to QUIC. The SOTA in Sisinni et al. (2023) also takes advantage of this characteristic to offer several mechanisms enabling connection migration from the server-side, such as Explicit and Pool of Address.

3.2.5. *Motivation*

As investigated in Sisinni et al. (2023), the native QUIC's mechanism and the server-side enhancement might enable the "connection migration" paradigm, but it is not yet a "smart" solution. On the other hand, Tan et al. (2020) was one of the first works to propose a smart connection migration to improve the network performance from the end user's perspective. However, the network infrastructure's dynamicity and complexity impact on the paradigm might have been left for future work by the community. The main contributors are the frequent infrastructure changes due to end-user mobility and unexpected events on the server side. Despite the effort to study smart connection migration in Tan et al. (2020), a single point of view from the client side or the running server itself might not be enough to reflect such network uncertainty and complexity toward the seamless network experience perspective. To be more precise, we argue that to fully exploit the smart connection migration's effectiveness, it is essential that the connection migration should also be triggered by a third-party entity, which has a global view of the infrastructure operational information. In the scope of this work, we focus on the first part of our suggestion, which is enabling a "third-party connection migration mechanism". Hence, in the following discussion, we will explain our solution design to cope with this matter.

3.3. Solution design

Taking a further step toward the suggested paradigm, we present our solution description in this section. First, we compare our sequence diagram of the connection migration process with the SOTA. We then propose a mechanism allowing the connection migration process to be issued from a third party on the client side. Following this, we explain our design decisions and the prospective impact.

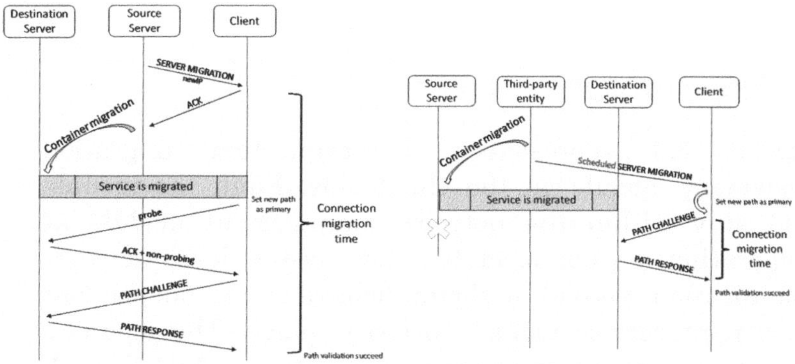

(a) The SOTA diagram (b) Third-party connection migration diagram

Figure 3.2. *Sequence diagram comparison. For a color version of this figure, see www.iste.co.uk/chu/softwaredefined.zip*

3.3.1. *Sequence diagram comparison*

Figure 3.2 illustrates the comparison in the sequence diagram between the SOTA and our proposed solution. Regarding the SOTA's approach, the client is only aware of the connection migration process once the SERVER MIGRATION (Sisinni et al. 2023) frame arrives from the source server. As soon as the client's ACK occurs, the container migration starts. Once the service is migrated, the client must send a PROBE message and validate the new path itself before it can use it. To ensure the success of the connection migration, the existing connection between the client and the source server must be maintained during the whole migration process.

On the other hand, taking advantage of the global overview from the third-party entity, our approach allows the container migration and the connection migration to be planned and executed before the connection migration starts. Instead of waiting for the container migration process to finish, our solution allows the client to use the new destination server as soon as the new IP address is available.

3.3.2. *System design*

Figure 3.3 represents our connection migration management module on the client side. First, the module should stay within the network management module at the operating system level to allow system-level network communication. Second, it should behave as an independent management service using a priority queue. The module's duty is to accept migration requests corresponding to each client's designated server and then to process those requests regarding priority. A notification mechanism is implemented to transfer the retrieved migration request from the system level queue to the QUIC connection's level and allow instant adoption of or a scheduled time frame for the new transferred IP as the new service destination.

A request migration's processing flow is described using the following numerical steps indicated in Figure 3.3:

1) The network interface *nic1* receives a connection migration request.

2) The network management module forwards the request to the connection migration module's queue.

3) The CM module will decide the request's processing order to execute the connection migration process regarding its priority and corresponding time frame.

4) The network management module sets the received IP as the new destination IP for the associated *connection-id*. Then, it executes intercommunication with the application as normal behavior.

5) After receiving the response from the application layer, the network management module passes the response to the physical network interface.

6) The connection migration is done, and the associated network interface *nic1* transmits the response's request using the new destination IP.

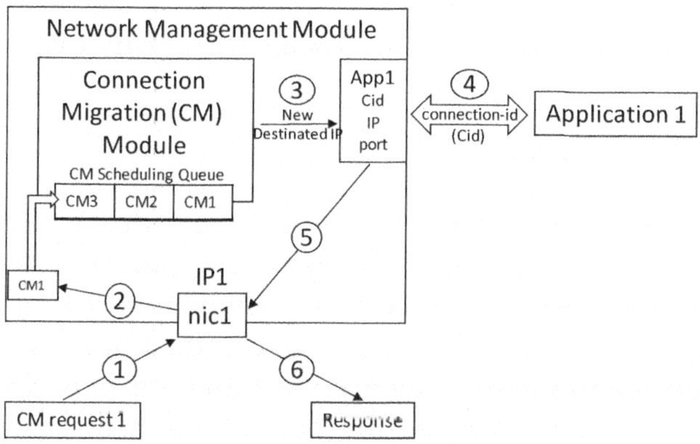

Figure 3.3. *Connection migration module's design and interactions*

3.3.3. *Prospective impact*

Compared to the SOTA approach, our most important advantage is optimizing the connection migration process. The essence of our approach derived from the service migration's final goal is toward a seamless network experience for the end-user. The longer the connection migration process, the higher the chance of an event in the infrastructure that requires a new migration. By giving the responsibility of the container migration process to the third-party actors and allowing the client to utilize the new destination IP immediately, the client is not required to take into account the complexity and dynamicity from the infrastructure's point of view. Conversely, the third-party entity, which can access all required information to make the service migration, will schedule the migration sequences.

This approach also eliminates the need to maintain the accessibility between the client and the existing server to allow the connection migration to proceed. The use case becomes quite useful when it is critical to maintain the original connection. By adopting the third-party entity, the connection could survive while a snapshot of the existing server is being deployed in the accessible infrastructure.

3.4. Performance evaluation

Having introduced our solution design and discussed the mechanism's benefits, we now compare our approach with the SOTA approach. First, we describe the technical setup used by the evaluation process. The details will be given in order: the experimental environment, metrics used and our experiment descriptions. Our obtained results and in-depth discussions conclude the section.

3.4.1. *Experimental environment*

Because of the advance in container technology and the source code published in Sisinni et al. (2023), reproducing the "connection migration" environment described in the SOTA required a reasonably small amount of effort.

Figure 3.4 indicates our setup: one VM as the client and the other two VMs as edge servers: the "source" server hosts the server container at the beginning of the experiment (*before* the migration occurs), while the "destination" server hosts the server container *after* the migration occurs. Each VM uses the exact same specifications: four vCPUs, 8 GB of RAM and 40 GB of disk, which runs Ubuntu 18.04.5 LTS. All the VMs join the same host-only network group supported by Virtual Box version 6.1.38. We chose the "connection migration time" as the performance metric: it is defined as the time the client sets the new IP as the primary address minus the new IP request's arrival time. We used two of the four introduced container migration types (cold and pre-copy) for our measurements. For each data point, we repeated the measurement three times. Since our proposed solution is related to the "connection migration" paradigm, further investigations about the container migration interactions are out of this work's scope.

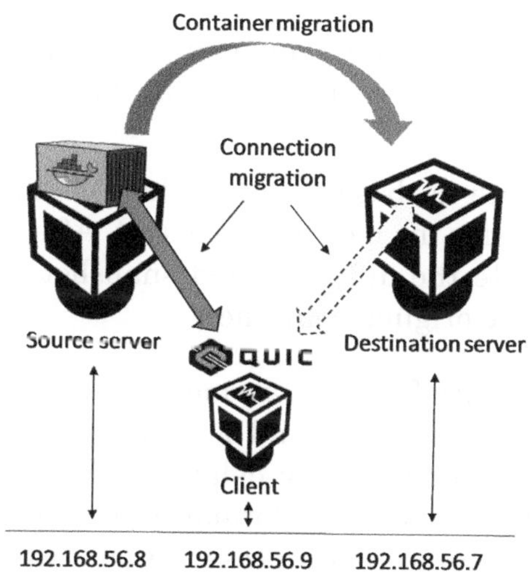

Figure 3.4. *Simulated experimental environment*

3.4.2. *Experiment scenario*

After deciding the required metrics and parameters, we evaluated our experiment. The experiment simulates the connection migration workflow in a typical scenario. While the client uses the service from the existing server (the source server), the client's network performance significantly decreases due to an unexpected event. The existing server finds no issue with its internal operation. Devoting assistance to improve the situation, the client should be served by another edge server candidate (the so-called "destination" server), expecting an improvement in its communication performance. Depending on the approach, while the SOTA takes advantage of the existing server, our solution uses a third-party server to trigger the connection migration process. Technically, the connection migration could only succeed after the container is migrated successfully. This study uses two basic container migration strategies for our measurements. The distinctions between the two strategies could be described as follows:

– cold: stop the container, create a snapshot of the current container, transfer to the new location and restore the snapshot;

– pre-copy: create a baseline snapshot of the current container and transfer the baseline snapshot. Afterward, stop the container, create the differentiated snapshot, transfer it and then restore the original container.

Using the above specifications, we measured the connection migration time using the SOTA and our third-party approach. First, we study the impact of the delay metric on the source server link, which causes the performance to decrease on the client side with the migration process. Second, we evaluated the impact of frequent migrations on the process. Such a scenario can typically happen in highly dynamic and changing conditions such as mobile environments. To this end, we measured the total connection migration time in multiple consecutive migrations to ensure our solution efficiency. An in-depth analysis will be presented in the following discussion.

3.4.3. *Influence of delay on the connection migration process*

Figure 3.5 illustrates the impact of the delay on the source server on the overall connection migration process inspired by our method versus the SOTA approach. As mentioned before, the work is our initiation into investigating the smart connection migration paradigm. Thus, only two fundamental container migration strategies are combined with selected connection migration candidates for our indication purposes. Despite a significant increase in the delay, the connection migration time in every combination increases slightly except for the server-side connection + pre-copy combination. In addition, the obtained results confirm that our solution significantly improves the connection migration time in both container migration strategies compared with the SOTA in each corresponding scenario. In our solution, the third-party entity executes the container migration before the connection migration proceeds. Hence, the client can finalize the connection migration process when the newly designated IP arrives. Therefore, our approach minimized the impact of the delay on the source server-side on the connection migration time.

3.4.4. *The influence of migration frequency on the connection migration time*

We now evaluate the influence of the migration frequency on the connection migration time. Based on the first result, we select the "cold" migration strategy as it offers the highest performance when combined with both connection migration approaches. The total connection migration time for four consecutive requests is plotted in Figure 3.6. We witness the same increasing trend in both connection migration approaches by the number of consecutive connection migration requests. However, our proposed solution shows a lower increasing slope, hence its overall performance is always better than the SOTA approach. The figure confirms again the improvement in connection migration time using our proposed approach. The trend also indicates our solution will continue to outperform the original approach as we increase the connection migration frequency.

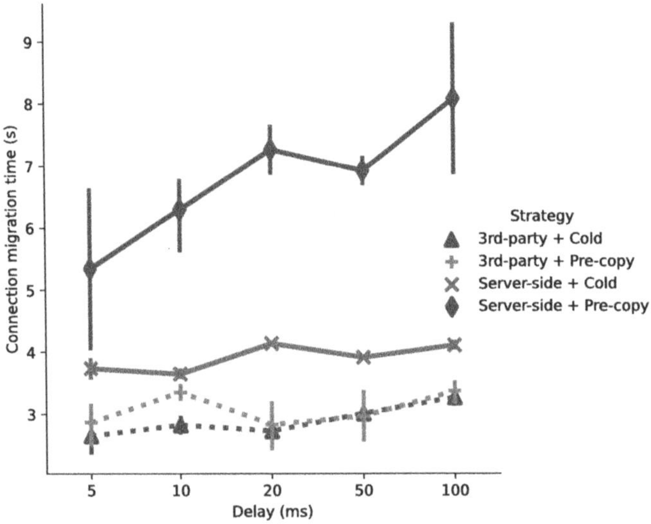

Figure 3.5. *Impact of the source server delay on connection migration time. For a color version of this figure, see www.iste.co.uk/chu/softwaredefined.zip*

3.5. Conclusion

This chapter presents an initial effort to enhance the SOTA toward "smart connection migration" regarding the context of this work. We propose a mechanism design and essential update implementation for the SOTA toward a "smart" connection migration. In our view, the intelligence-based perspective has opened the door to employing the full capacity of the "connection migration" terminology in various domains, especially in high QoS requirement environments. In the H-IoT landscape, especially in the healthcare transport scenario, the paradigm is expected to allow the end-user to experience the lowest latency to the edge servers and a seamless network experience despite infrastructure changes to satisfy its strict communication requirements. We also compare our solution's performance with the SOTA using the connection migration time metric in the equivalent simulated environment. Our results confirm our solution's efficiency (50% faster than the original approach in connection migration time) and enable further investigation into this interesting research direction. Despite the significant improvement in performance metrics, we plan to study the "smart connection migration" paradigm in the fully-fledged aspects with more insightful metrics in our future work.

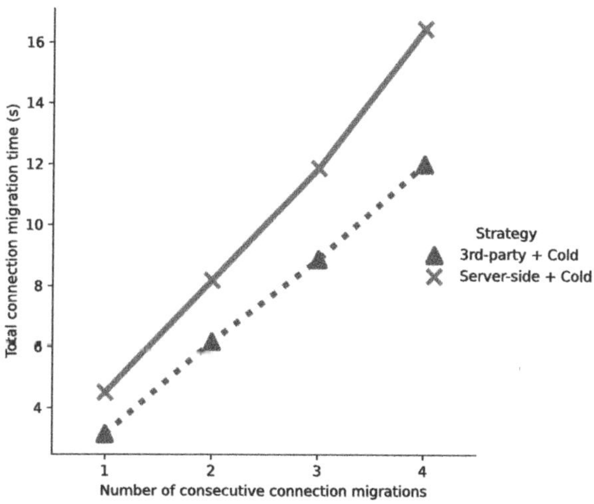

Figure 3.6. *Impact of consecutive connection migration on connection migration time. For a color version of this figure, see www.iste.co.uk/chu/softwaredefined.zip*

The next chapter will present a proven solution that works independently from the SDI's influence. The solution aims to assist the end-user in achieving maximum network performance by taking advantage of existing multiple network interfaces for healthcare emergency communications.

Chapter 4

Generic Adaptive Deep Learning-based Multipath Scheduler Selector

Predicting the future isn't magic, it's artificial intelligence

Dave Waters

This chapter provides a proven solution to enhance end-users network capability in highly dynamic and heterogeneous environments with uncertain SDI reachability, particularly in a healthcare transport environment.

4.1. Introduction

High bandwidth, guaranteed reliability, and ultra-low latency are among the staggering promises of 5G, and mobile end-users are eager to take advantage of its worldwide deployment. The services that benefit the most from this enhancement could be basic users with their ever-growing network requirements and, more critically, healthcare applications and autonomous vehicular networks. These use cases have a common requirement of much higher network capabilities when compared to traditional applications. Simultaneously, the emergence of multipath-enabled protocols such as MP-TCP and MP-QUIC has offered an elegant solution to exploit this tremendous network power. Taking advantage of multiple available network interfaces at the same time, these transport protocols allow the end-user device to have better throughput and lower latency, as well as a fallback solution in case of a network failure. The identified characteristics emphasize the necessity of adopting multiple network interfaces in high-QoS scenarios, particularly within the context of this work, which explores solutions for H-IoT emergency situations. Figure 4.1 indicates the third contribution prospect as an independent solution in the thesis landscape.

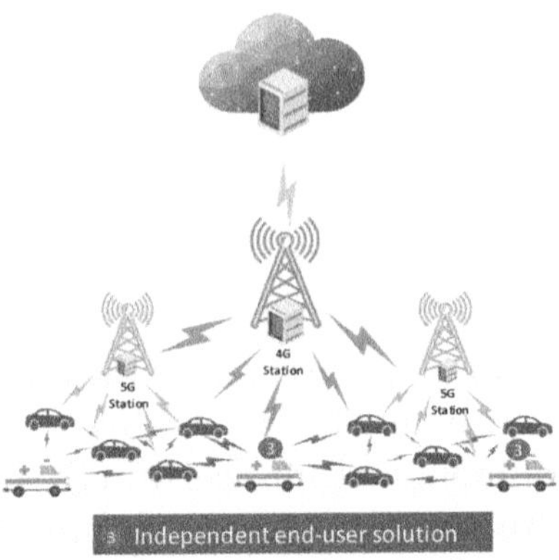

Figure 4.1. *Third contribution prospect in the overall framework. For a color version of this figure, see www.iste.co.uk/chu/softwaredefined.zip*

In addition, the 3rd Generation Partnership Project (3GPP) has also emphasized this potential cooperation in the Technical Specification (TS) 23.501 (Release 16) by claiming that the Access Traffic Steering, Switching, and Splitting (ATSSS) architecture could be enhanced by multipath transport protocols (3GPP 03.2020). One of the most attractive research directions in the community is the multipath scheduler, a piece of software that decides which network interface to use at a given time. The scheduler plays a central role as it affects the overall network performance. Looking at the recently developed process of the multipath scheduler, an obvious trend could be witnessed. Most cutting-edge scheduler proposals aim at competing in drastically asymmetric network conditions. To simplify, the environment in which they conduct their evaluation systematically has a “good” network connection and a “bad” network connection, most of the time with much lower link quality. However, this scenario becomes less and less realistic, considering the advancement of communication technologies, such as 5G wide deployment, WiFi 6 and multiple cell usage technologies. This fact leads to another concern for the researchers: Should it be a more generic solution, which takes advantage of both old-fashioned and the most cutting-edge solutions to provide the best performance for the end users, especially in high dynamicity and fluctuation?

To our knowledge, six schedulers have made their reputation in the community in the last few years: round-robin, minRTT, BLEST, ECF, RELeS and Peekaboo. While Peekaboo, the state-of-the-art scheduler, seems to provide the best performance in some specific environments, there is some evidence that other existing schedulers may outperform it when facing more realistic network conditions. Based on this observation, we proposed GADaM (Chu et al. 2022), a scheduler *selector* whose role is to select the most appropriate scheduler under specified network conditions dynamically. Instead of using a deterministic approach to select a scheduler and hope for the best performance, the scheduler selection process is dynamic and adaptive, taking advantage of previous performance observations in the equivalent network conditions.

The contributions of this chapter are as follows:

– introducing the "scheduler selector" paradigm;

– a prototype of GADaM, a deep learning-based instance of a "scheduler selector", for an Android environment;

– validating the paradigm's efficiency in both a simulated environment and three real-world scenarios in an urban area near Paris, France.

4.2. Related work

In this section, we present the recent multipath transport protocols and state-of-the-art multipath schedulers. Then, we review the performance of these multipath schedulers over different levels of path dynamicity and discuss their limitations.

4.2.1. *Multipath transport protocols*

For decades, TCP and UDP have been the only transport protocols in the pervasive TCP/IP stack, although they were built with the intrinsic assumption that only a single path may exist between two end-hosts at any given time. A quick look at the plethora of network interfaces that equip today's end-hosts shows how wrong this assumption has become. Therefore, multipath transport protocols have been developed to enable the simultaneous use of various paths that may exist between a sender and a receiver. Any such protocol can bundle different network interfaces and switch between them transparently. The most popular implementations to date are Multipath TCP (MPTCP Ford et al. (2013)), Concurrent Multipath Transfer for SCTP (CMT-SCTP Iyengar et al. (2006)) and Multipath QUIC (MPQUIC De Coninck and Bonaventure (2017)). In order to make the most of the multipath approach, research focuses on three orthogonal aspects: congestion control, path management and scheduling. Indeed, the multipath scheduler plays a key role as it is responsible for distributing data over multiple paths. Poor scheduling decisions can introduce Head-of-Line Blocking (HoLB) or increase out-of-order packet arrival especially when the paths are heterogeneous. The next sections will take a deeper look at the design principles of multipath scheduling algorithms.

4.2.2. *Multipath scheduling algorithms*

A multipath scheduling algorithm aims to make the best use of the available paths by choosing which path should be used to transmit the next packet in order to avoid congestion in the multipath environment. It is intended to address certain problems on the network, in particular, out-of-order arrivals, head-end blocking, etc. When dissecting current multipath schedulers, we observe that they can be split into two categories: those based on predefined rules, and those based on machine learning (ML).

4.2.2.1. *Non-ML-based multipath scheduling algorithms*

One of the most basic schedulers is the round-robin (RR) scheduling algorithm, which sends data over each available path sequentially as long as there is space in the congestion window (CWND). Since it does not rely on any information specific to the path characteristics to make its decisions, it usually results in a poor performance, especially in a heterogeneous environment. The second basic scheduler is the minRTT scheduling algorithm, which is based solely on the round trip time (RTT). Indeed, after determining the RTT and after checking the space in the CWND, it sends the packet on the path that has the lowest RTT. Currently, it is the default algorithm in MPTCP (Ford et al. 2013) and MPQUIC (De Coninck and Bonaventure 2017). The Blocking Estimation Scheduler (BLEST) is another scheduler based on predefined rules which has the particularity of offering a waiting mechanism to delay the emission if only the path with the highest RTT is available. BLEST can thus decide to accept the lowest (Ferlin et al. 2016). In the same spirit as BLEST, the ECF (Earliest Completion First) algorithm also applies a waiting mechanism to estimate the idle time of the lowest RTT path and wait to transmit on this path (Lim et al. 2017). RAVEN is another MPTCP scheduler that uses redundant communication across multiple networks based on RTT measurement and RTT prediction for future communications, with the aim of reducing application interaction delay and hiding high queuing latency (Lee et al. 2018). Several other schedulers based on predefined rules must be mentioned, such as MuSher (Aggarwal et al. 2022), an Agile Multipath-TCP Scheduler for Dual-Band 802.11ad/ac Wireless LANs or Slide Together Multipath Scheduler (STMS) (Shi et al. 2018), which aims to reduce out of order packet arrivals. Finally, we can also mention estimated transfer completion time (ETC), which offers a scheduling mechanism that assigns data to paths with transfer simulation (Zeng et al. 2022).

4.2.2.2. *ML-based multipath scheduling algorithms*

To address the performance limitations of rule-based schedulers, we notice a growing interest in the development of ML-based multipath schedulers. ReLeS uses a neural adaptive multipath scheduler for MPTCP based on deep reinforcement learning (DRL) (Zhang et al. 2019). Thus, ReLeS applies deep Q-network (DQN) to generate the control policy for packet scheduling from online data. Peekaboo (Wu et al. 2020) is another learning-based scheduler that first selects a deterministic strategy based on Linear Upper Confidence Bound (LinUCB) and then applies a stochastic adjustment strategy applied for the MPQUIC protocol. Modified-Peekaboo (M-Peekaboo) extends the Peekaboo learning scheme to 5G mmWave networks for the path selection (Wu et al. 2021b). GADaM offers another approach by positioning itself as a scheduler selector (Chu et al. 2022). After an analysis that, showed that, in the same scenario with varying network conditions, none of the aforementioned schedulers could systematically present the best performance, the authors therefore proposed a deep-learning-based approach that consists of selecting the most adapted scheduler, given the network conditions observed in a previous time window. FALCON is another multipath scheduler that has the specificity of combining the benefits of both offline and online learning (Wu et al. 2022). The solution optimizes both the adaptation speed and accuracy based on network conditions. Lee and Yoo (2022) proposed a scheduler based on deep reinforcement learning dedicated to multimedia streaming. In this proposal, both the delay and throughput are considered as a reward for reinforcement learning.

4.2.3. *Multipath scheduling performance over heterogeneous paths*

In the various research works, we observe that the experimental validation of the various schedulers is mainly carried out in simulated and static environments. Moreover, these validations are most often carried out in networks with low heterogeneity and for a fairly low volume of data exchange. If we neglect these various shortcomings, we see a clear superiority of ML-based solutions.

Real-world experimental evaluations are more the exception than the rule. In Paasch et al. (2014), the authors proposed a solution based on MPTCP to measure performance in both simulated (Mininet) and actual (NorNet testbed) environments. In general, real-world evaluations follow the path paved by De Coninck and Bonaventure (2019), who developed a multipath-enabled mobile application for testing multipath performance over MPTCP and MPQUIC. For example, this is what Wu et al. (2020) have achieved in their work, with the most complete set of validations, to the best of our knowledge. Nevertheless, they clearly express a possible limit of Peekaboo in mobile scenarios, particularly when network conditions change drastically and frequently. However, they did not verify the behavior of their solutions in such a context. The mobile validation environment in Wu et al. (2021a) uses real-world network traces but in an emulated environment.

4.3. Prototype design

After a brief discussion about the related work, we will describe the scheduler selector paradigm in this section. The GADaM prototype design as well as a proposed architecture to implement the paradigm in the network stack will also be presented.

4.3.1. *Scheduler selector paradigm*

Taking the network condition of all the paths as the input, a pre-trained model is used to select the most suitable scheduler to achieve the best network performance, hence the novel paradigm "scheduler selector" is invented. The scheduler selector works as a classifier because the model tries to identify the most suitable scheduler for a type of network condition. Figure 4.2 describes the workflow of the proposed paradigm.

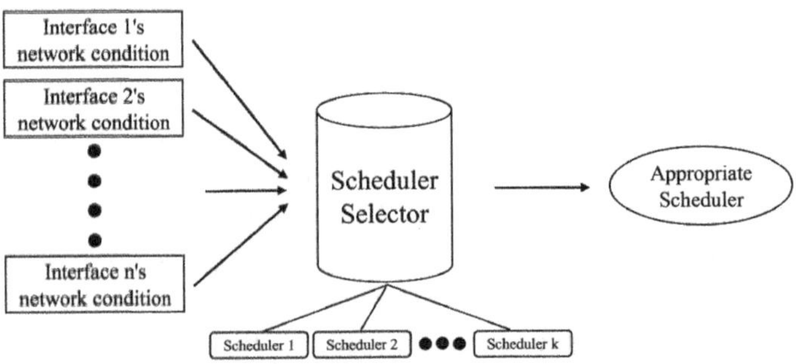

Figure 4.2. *The scheduler selector*

4.3.2. *Prototype design*

In this section, the scheduler selector's design process will be discussed as follows: the input, the output and the model.

4.3.2.1. *The input*

The input for the model consists of the network conditions of multiple interfaces. A network condition of an interface is defined as an identical status of the data transmission line. These lines' status is usually measured by the network metrics (Yang et al. 2018). Because the target for the improvement is positioned in the transport protocol, it is more feasible to have access to the transport and network metrics. Hence, such metrics are prioritized to become the representative metrics for the network condition in this work, which are packet loss ratio, network delay and bandwidth. In the proposed prototype, these three metrics are used to describe the network condition of an interface because of the emulator's support. However, only two interfaces are used to build the prototype to create the most similar environment to the current research work, hence the input is constructed from six features: first interface's bandwidth, second interface's bandwidth, first interface's delay, second interface's delay, first interface's packet loss and second interface's packet loss.

4.3.2.2. *The output*

The output of the scheduler selector would be the label of the suggested scheduler. The suggested scheduler is the fairest scheduler, which achieved the highest performance in the benchmark between all the schedulers in that particular multi-interface network condition. To build as generic a prototype as possible, most of the state-of-the-art schedulers are used. The list of the available schedulers is as follows: round-robin, minRTT, BLEST, ECF and Peekaboo. ReLeS is excluded from this work because its implementation is not available.

4.3.2.3. *The model*

Since the addressed problem is a classification problem, two well-known approaches are chosen: traditional ML and deep learning-based approach. Two of the feasible candidates selected to become the baseline for this work are the K nearest neighbors (KNN) algorithm and the random forests algorithm along with a proposed deep learning architecture.

A dataset is synthesized as a csv file for training and testing the models. Each row of this file is a feature vector and its label. The feature vector is described in section 4.3.2.1 and the label is described in section 4.3.2.2. This dataset will be used for training and testing the model before deploying in the practical stack.

4.3.2.4. *Deep learning architecture*

The second contribution of the chapter is the proposed deep learning architecture for the prototype, which will then be discussed separately. The dataset's structure is fairly simple and does not possess any special characteristic, so the normal deep neural network architecture with feed forward function is chosen. The proposed neural network architecture is described in Figure 4.3.

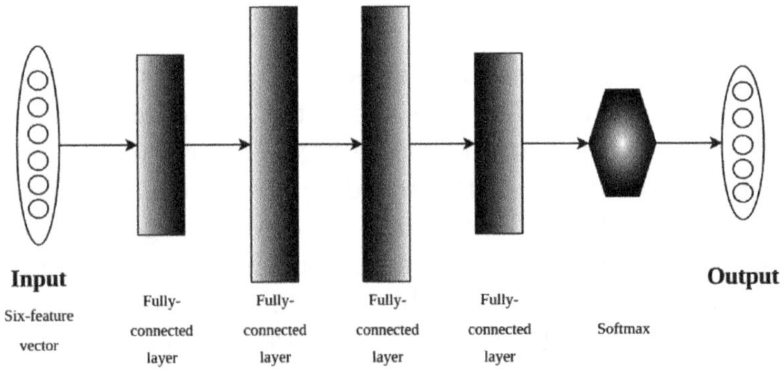

Figure 4.3. *Proposed deep neural network architecture*

Figure 4.4 indicates one of the suggested architectures to implement the proposal paradigm "scheduler selector" in the practical network stack. Taking the input as the network condition, according to each request from the client side, the appropriate scheduler will be suggested to the server by the "scheduler selector".

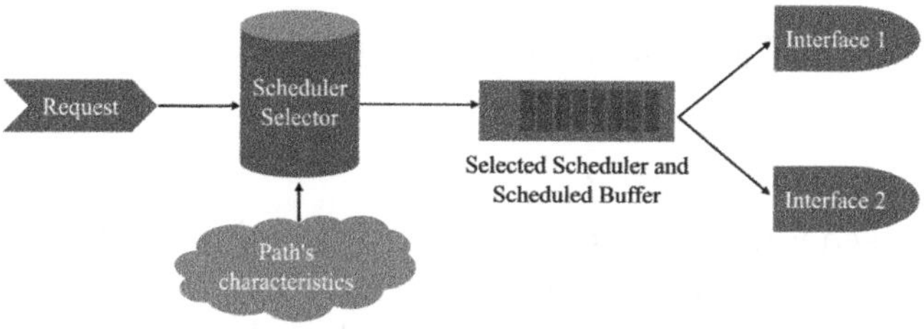

Figure 4.4. *Proposal architecture. For a color version of this figure, see www.iste.co.uk/chu/softwaredefined.zip*

4.4. Simulated evaluation

The experiment process as well as the performance evaluation of each approach in a simulated environment is explained in this section. Section 4.4.1 discusses the experiment environment. The validation process of the performance differentiation of each scheduler in various network conditions will be presented in section 4.4.2. Technical details about performance of the classification approaches is presented in sections 4.4.3 and 4.4.4.

4.4.1. *Experiment setup*

4.4.1.1. *Network topology*

A simple topology composed of a server, a client, a switch and two distinguish connections is used to indicate common multipath network topology. As indicated in Figure 4.5, the topology used is the standard in this research topic.

The performance data of each multipath scheduler in a heterogeneous environment will be collected using the exact same topology except for the variation in network link quality.

Figure 4.5. *Experiment topology*

4.4.1.2. *Experiment scenario*

For each scheduler, the experiment is conducted with the following scenarios. For each network condition, the client will send a request to the server asking to download a 2 Mb file 15 times. The server will use the specified scheduler, and then log the completion time for each request. The completion time is the metric for bench-marking the scheduler performance. The lower the completion time is, the better the scheduler.

4.4.1.3. *Path's characteristic metrics*

To overcome the infinite searching space of each of the network metrics, each metric, partially based on previous work such as Zhang et al. (2019) and Wu et al. (2020), is going to be sampled in a range. The more steps that are chosen, the more experiments that need to be conducted; then, the exponential function combined with results from previous works (Ferlin et al. 2016; Lim et al. 2017; Zhang et al. 2019; Wu et al. 2020) is used to generate the metric steps in Table 4.1.

Bandwidth (Mb)	Delay (ms)	Packet loss (%)
4, 8, 16, 32, 64	1, 4, 16, 64	1, 2, 4

Table 4.1. *Path's characteristic metric*

4.4.1.4. *Tools and framework*

The framework for the prototype implementation is based on MP-QUIC, and the reference implementation of multipath schedulers (RR, minRTT, ECF, BLEST, Peekaboo) is used from Peekaboo's published source code. ReLeS is mentioned frequently in this work; however, the implementation has not been published, so the result was not conducted.

The reason for choosing MP-QUIC for this prototype is that QUIC is a cutting-edge protocol implemented in a user space, allowing researchers to modify, develop and test quickly.

For emulating the network topology, Mininet (Handigol et al. 2012) is used. Tc Netem (Hemminger 2005) is used to control the link quality diversity. The implementations for the ML algorithms are from the Scikit-learn (Pedregosa et al. 2011) package. The deep learning framework used in this prototype is Pytorch (Paszke et al. 2019).

4.4.1.5. *Hardware specification*

The experiments are conducted on a laptop with an i5-8650u CPU, 16 GB of RAM, with Ubuntu 18.04 LTS, 256 GB solid-state disk. The server and the client is emulated with the same virtual configuration: 1 vCPU, 3 GB of RAM, 10 GB solid-state disk.

4.4.2. *Initial dataset analysis*

By ranking the downloading completion time of each algorithm for each network condition, the result is indicated in Figures 4.6 and 4.7.

Figure 4.6 is a histogram of the number of times which algorithm performs the best. The benchmark is based on the downloading completion time according to each scheduler. The lower the completion time, the higher the performance the scheduler has achieved. The distribution is nearly equal between the algorithms. Even the most primitive scheduler, such as round-robin, possessed a significant highest performance number. This means each scheduler provides the best performance in some scenarios but not all. Figure 4.7 indicates the number of times each algorithm performed the worst. For example, Peekaboo might achieve a significant result when comparing itself with other algorithms in Figure 4.6, but in Figure 4.7, it possessed the second-largest number of times performing the worst.

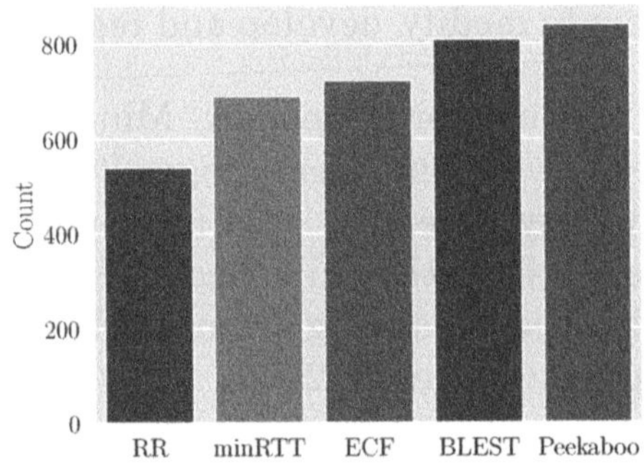

Figure 4.6. *Best performance histogram. For a color version of this figure, see www.iste.co.uk/chu/softwaredefined.zip*

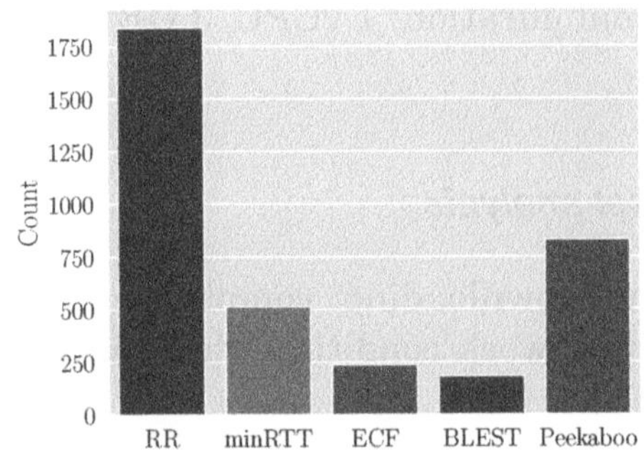

Figure 4.7. *Worst-performance histogram. For a color version of this figure, see www.iste.co.uk/chu/softwaredefined.zip*

The initial dataset analysis has confirmed there is no best scheduler for all the scenarios. Hence, extra work has been conducted to verify the classification performance of the suggested approaches to find the most suitable model. The dataset is split into three subsets: the training dataset, the validation dataset and the testing dataset. The distribution of classes in the training, validation and testing dataset are plotted in Figure 4.8 to make sure all the classes are split equally among the dataset. The ratio for splitting the data is 72% for the training dataset, 8% for the validation dataset and 20% for the testing dataset.

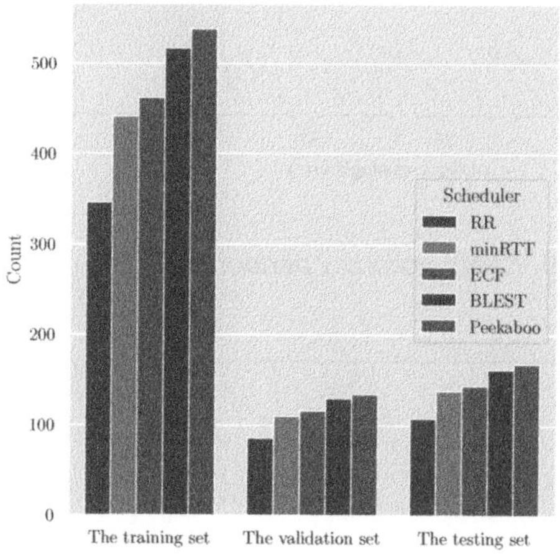

Figure 4.8. *Class distribution in the training, validation and testing set. For a color version of this figure, see www.iste.co.uk/ chu/softwaredefined.zip*

4.4.3. *Traditional machine learning's performance evaluation*

The first proposed approach is the traditional ML approach. The two well-known candidates selected for the experiment are KNN and random forest. The performance evaluation is discussed as follows. For the KNN algorithm, the number of neighbors starts from 1 to 50. For random forests, the number of the estimator is from 10 to 100. Figures 4.9 and 4.10 describe the classification's accuracy of these algorithms corresponding to their hyperparameters.

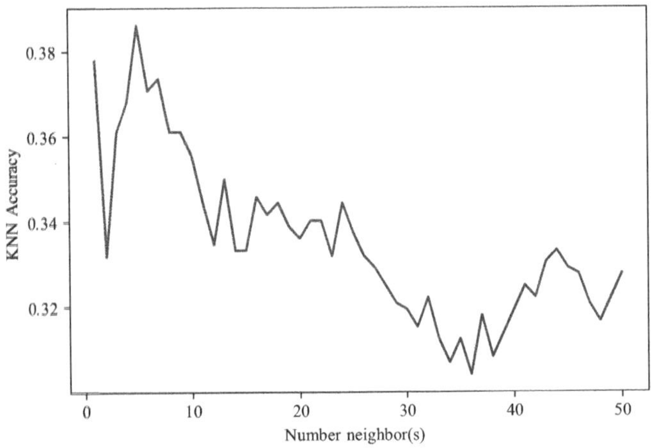

Figure 4.9. *KNN's accuracy performance*

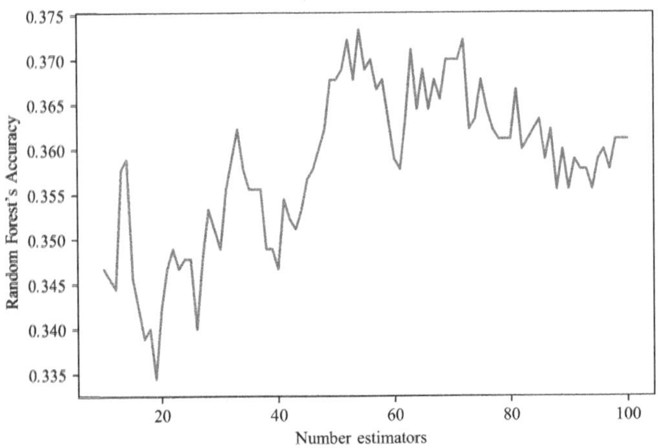

Figure 4.10. *Random forest's accuracy performance*

The traditional approaches performance fluctuates and is unpredictable. The most significant accuracy in the training phase of both algorithms is not higher than 40%, which is too low for evaluation. This means these approaches are not suitable for practical use because they could not effectively select the appropriate scheduler for each network state; hence, reasonable run-time and QoS improvement could not be achieved as the proposal purpose. This confirms that the traditional ML algorithms could not learn the model from the data.

4.4.4. *Deep learning's performance evaluation*

Subsequently, the second proposed approach and the proposed DFF architecture is evaluated. The analysis is as follows. The training and validation accuracy during the training phase of the neural network is plotted in Figure 4.11. The value of the loss function on the training and validation set is indicated in Figure 4.12. The accuracy of the model gradually increases with each epoch while the loss function value also steady decreases. This indicates the model was successful in learning the identical detail of the network condition for each scheduler. To have the neural network converge, the training phase needs around 4,000 epochs to achieve more than 90% of accuracy on the training set and validation set.

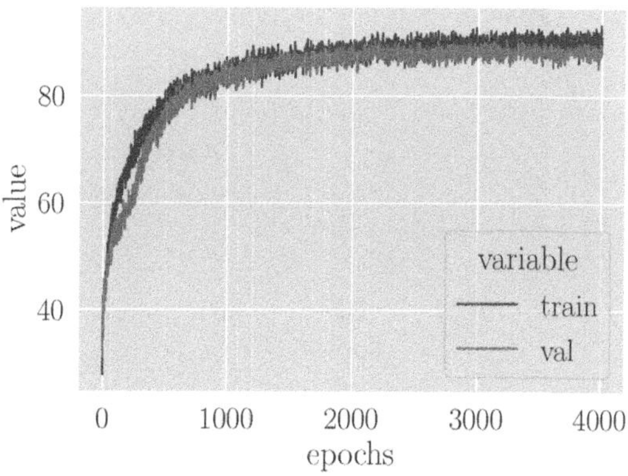

Figure 4.11. *Accuracy on the training and validation set. For a color version of this figure, see www.iste.co.uk/chu/softwaredefined.zip*

The model performance could be investigated deeper with the confusion matrix of the testing set, as shown in Figure 4.13. Each scheduler is labeled as follows: 0: round-robin, 1: minRTT, 2: ECF, 3: BLEST, and 4: Peekaboo. The classification performance of the deep learning approach completely overtakes the traditional approaches with more than 95% accuracy on the training and over 91% on the testing dataset. The confident interval (CI) of the model for the testing set with 95% could be calculated using the following formula:

$$CI = error \pm const * \sqrt{\frac{error * (1 - error)}{n}}$$

The error according to the testing set accuracy is 8.2%. The constant for 95% confidence is 1.96. The validation set's size is 8% of the total dataset. The number of records of the dataset is 3,600, so the number of records is 288. Hence, the CI for the testing set with 95% is:

$91.8 \pm 3.16\%$

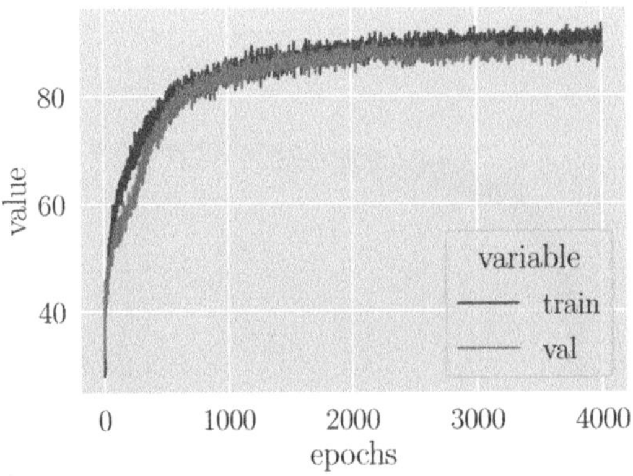

Figure 4.12. *Loss value on the training and validation set. For a color version of this figure, see www.iste.co.uk/chu/softwaredefined.zip*

Because the approach proposed is offline learning, its runtime performance is expected to not affect the overall performance of the network stack. It is practical to apply the scheduler selector in the network stack of the protocol with the proposed deployment architecture, as shown in Figure 4.4.

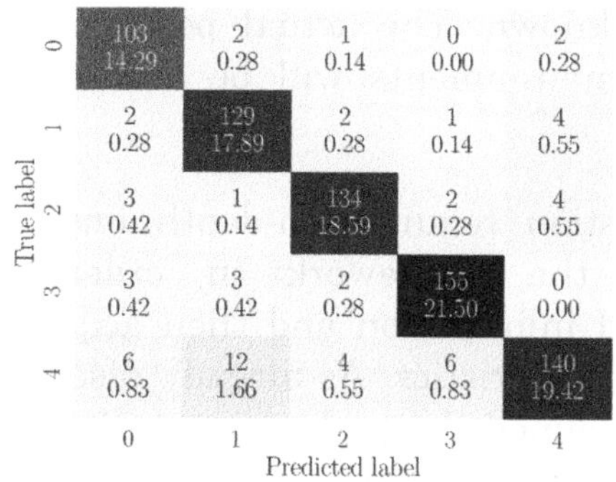

Figure 4.13. *Confusion matrix of the testing set*

4.5. Practical evaluation

After measuring GADaM's performance in the deep-learning accuracy metric in a simulated environment, we evaluate GADaM's actual performance in realistic environments. Therefore, the technical aspects related to this evaluation will be described. We start with a brief recap of the design process of GADaM's practical evaluation. Finally, we describe the experiment environment and setup.

4.5.1. *Brief overview of GADaM*

GADaM's guiding principle is to select the best scheduler from a series of observations of the current network conditions.

This raises two mandatory questions:

– What is the best scheduler for the current network conditions?

– How can we measure the current network conditions?

In our first work (Chu et al. 2022), we partially solved this problem by proposing a deep-learning-based model to select the best scheduler for network conditions over two paths. However, the effectiveness of this model in a real network remains largely unknown. The actual performance of this model in various environments will be discussed later in sections 4.5.4–4.5.8.

The second question required in-depth implementation involvement with the framework in order to access transmission-related information and then acquire network metrics as path characteristics. As already discussed in our previous work, we defined the path characteristic consisting of three network parameters: delay, loss rate and bandwidth. The following discussion outlines the method for measuring network metrics before they are processed by GADaM to generate the most suitable scheduling recommendation.

4.5.1.1. *Delay*

The delay of a network connection is defined as the time required for the data to travel from the source to the destination. To acquire an estimation of the delay in the most simple way, we divided by 2 the RTT, which is measured and smoothed by the congestion control module inside the framework.

4.5.1.2. *Bandwidth*

The definition of bandwidth is the amount of data that can be sent over a connection during a period of time, usually calculated in megabytes per second (Mbps). Taking advantage of the ACK mechanism within the framework, we measured the time to receive the ACK from a packet, then took the packet size and divided it by half of the ACK time. The ACK time here is equivalent to the RTT; however, it is related to the programming technique, and using this mechanism helps us to speed up the implementation process. Taking the precaution of ACK loss that might lead to undefined data sending time, we also apply a slicing window technique to eliminate outliers.

4.5.1.3. *Loss rate*

Loss rate, as the name suggests, is the percentage between the packets that are sent by the client but not received successfully and the total sent packets within a duration of time. Receiving ACK is one of two main parts of measuring packet loss. The congestion control module in the original code base provides a convenient way for counting ACK packets. In our framework, we make use of this mechanism and also add a slicing window with a pre-defined timeslot to calculate the packet loss rate.

In this prototype, we implement GADaM's scheduler suggestion mechanism that would be provided on a per packet basis. Taking the paths' characteristics as input, for each packet, GADaM will suggest the most suitable scheduler based on its prior knowledge and use that scheduler to take the next scheduling action.

To confirm that our model works as expected, we take a brief look at the GADaM internal decisions in a typical experiment run and plot them in Figure 4.14. The bottom plot represents the scheduling decisions made by GADaM over a 5,000 packet period, while the top plots represent zooms over two particularly interesting periods. The Y axis of the three plots indicates the name of the selected scheduler from among five candidates. As we can see, a different input (i.e. network conditions) leads to a different scheduler selection.

The figure shows GADaM's ability to make different decisions on a per-packet basis. For instance, in the top-left plot, GADaM rapidly switches from RR to Peekaboo, then minRTT, according to the measured network conditions. As expected, steady network conditions lead to the same decision, hence the long period in the bottom plot (from 2,500 to 4,000) where a scheduler change would instead decrease the performances.

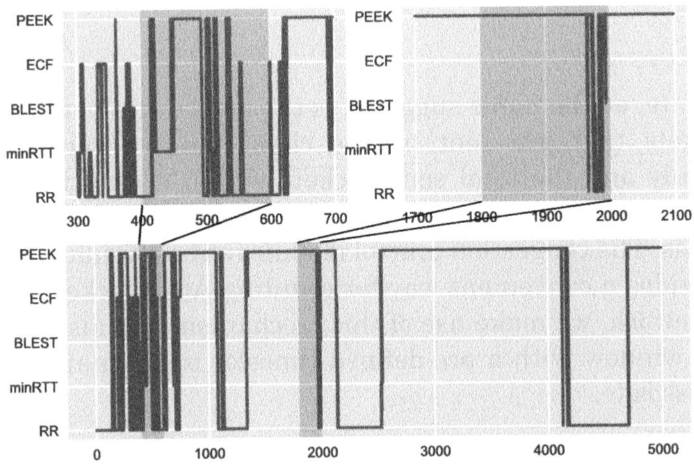

Figure 4.14. *GADaM's prototype scheduler suggestion mechanism*

4.5.2. *Extensible modular scheduler evaluating framework*

To evaluate GADaM's performances in a real environment, one must take mobility and fairness into consideration. Therefore, we improved the MP-QUIC implementation provided by Wu et al. (2020) to build an extensible modular scheduler evaluation framework. While the very first version was implemented by De Coninck and Bonaventure (2017) in Go Lang, there is no ready-to-use version to deploy on Android and iOS. Such a multipath-enabled framework for a mobile phone is nevertheless required to trigger experiments on these well-known mobile OS. However, De Coninck and Bonaventure (2017) provided a multipath tester application for iOS 11. Regarding the state of the scheduler implementation, Wu et al. (2020) have implemented some popular schedulers in their implementation. Considering all of these facts, we decided to take advantage of the existing implementation in Go Lang as a core element to achieve the fastest integration speed. Our contributions to this framework to enable GADaM evaluation in practice are as follows:

– add the modular evaluating procedure for both existing and potential schedulers;

– integrate the workflow for the Android deployment of MP-QUIC client from the original source code;

– implement the estimation of the path characteristics into the framework;

– automate the experiment orchestration to ensure simultaneity and therefore identical network conditions;

– collect and filter the results, and produce extensive plots automatically.

In frequent dynamic and fluctuating environment scenarios like healthcare emergencies, especially in highly dense urban areas, packet loss and retransmissions are expected to happen occasionally. The other usual metrics, such as throughput and jitter, might not indicate the overall system's performance. We use the download file scenario to evaluate the scheduler's performance, which has been used for years in this kind of work (Wu et al. 2020) as a standard metric. In addition to the software extension we made to allow us to conduct our experiment, a hardware infrastructure is also required to be deployed to measure the download completion time.

The infrastructure is composed of three main parts: a client cluster, a server cluster and the network infrastructure. The deployment of these elements could be found in the discussion of our experiment process in sections 4.5.4 and 4.5.5.

4.5.3. *Experiment environment*

In order to fully understand the real performance of each scheduler in multiple actual environments, we conducted our experiments on a set of bare-metal equipment in both static and mobile environments. Their characteristics are summarized in the following sections. A complete description of these environments can be found in Diakhate et al. (2022).

Environment	Technology	Bandwidth variation (%)	RTT variation (%)	Loss (%)
Static in-house	WiFi	30	6.2	0
	LTE	0.6	0.9	0
Mobile metro line	WiFi	44	4	10
	LTE	44	16	0
Mobile in-vehicle	WiFi	30	27.5	11.5
	LTE	10	2.8	62

Table 4.2. *Variations of networks conditions*

4.5.3.1. *Static environment*

The static environment we chose for the experiment is an in-house environment. Therefore, we do not expect GADaM's improvement to be as significant as in more dynamic environments. We used this environment as a baseline to validate GADaM's ability to made coherent, stable decisions.

The multipath network infrastructure will be provided by two lines, the first one is a 4G signal by a cellular operator, and the second one is a WiFi signal by a fiber line. During the experiment process, the mobile phone will not move at all. Besides, even in a static environment such as in a house, office or library, the network condition could still fluctuate since the infrastructure might be affected by countless factors then have a negative effect on the end-user device performance.

4.5.3.2. *Mobile environment*

We therefore tried our best to conduct experiments in multiple urban mobile scenarios in order to pinpoint the full potential of GADaM in dynamic scenarios. For the sake of completeness, we decide to run our experiment in two usual urban environments: the public metro transportation and a private in-vehicle transportation.

The system architecture used to conduct the experiment is illustrated in Figure 4.22, and is absolutely identical to the framework we designed in Diakhate et al. (2022). It uses five smartphones as MPQUIC clients, four servers (hosted in a cloud in the United States), each running a particular scheduler, and a fifth server running GADaM. Each client uses a specific Android application to download a file hosted on a predefined server.

The description of each experiment could be found in section 4.5.5.

For the first mobile environment, to avoid interruption during the acquisition process, the subway route with the longest distance of 23 km of the Paris metro is selected. The journey requires 53 min to accomplish from the first departure to the final destination. The movement will be controlled by a driver and will stop 38 times according to this line's schedule.

The selected route on the map in both directions is illustrated in Figure 4.15.

For the second mobile environment, which is in-vehicle transportation, we decided to use a personal vehicle to be more in control of the movement. During the experiment process, the vehicle's speed varies between 30 and 50 km/h. The stopping frequency depends on the traffic signs, which is occasionally. We conduct our experiment during peak hours with the aim to have a high urban density.

At the beginning of our experiment, to ensure if a further experiment is necessary, we decided to conduct a trial experiment, which require much less effort and time, at first. After retrieving some promising results from our trial run, we proceed on conducting our actual experiment with the maximum scheduler candidate capacity to help us conclude our studies. The description of each experiment could be found in sections 4.5.4 and 4.5.5.

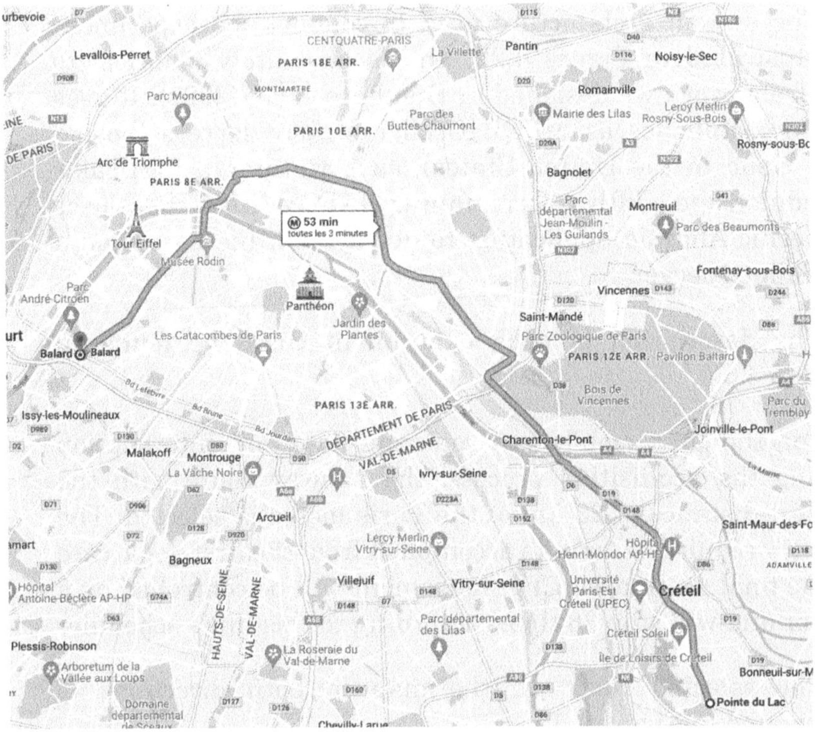

Figure 4.15. *Selected metro line for the first mobile scenario. For a color version of this figure, see www.iste.co.uk/chu/softwaredefined.zip*

4.5.4. *Trial run*

During this trial run, only two mobile phones and two servers will be used for measurement. Since our target is GADaM evaluation, we specified a dedicated mobile phone and a dedicated server for GADaM's usage. The other ones would play the same role for other scheduler candidates, only one at a time. This trial run experiment was conducted in a static scenario. After our initial analysis on the collected dataset, we found some promising insights. Before proceeding to the actual experiment result and discussion, the following section discusses the trial run process and results.

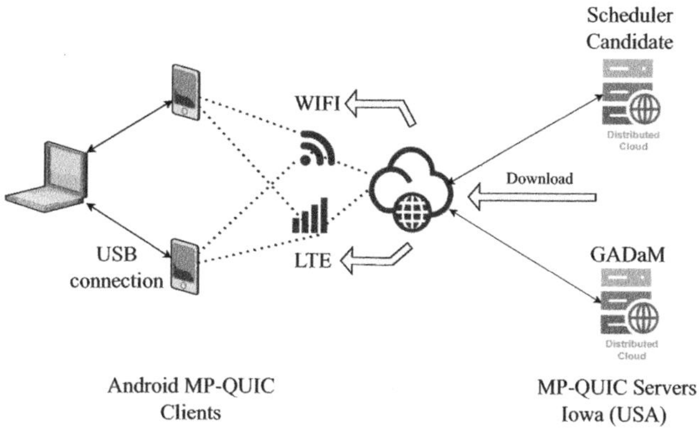

Figure 4.16. *Trial run architecture. For a color version of this figure, see www.iste.co.uk/chu/softwaredefined.zip*

4.5.4.1. *System architecture*

The experimental hardware setup for this trial run is in static environment and illustrated in Figure 4.16. As mentioned before, in this trial run, only one scheduler candidate will be compared to GADaM's prototype implementation at a time using the extensible scheduler assessment framework. Figures 4.17 to 4.21 illustrate the performance comparison between round-robin, minRTT, BLEST, ECF, Peekaboo and GADaM's prototype implementation using download completion time with different file sizes as metrics. Each experiment is conducted in the same time frame for each scheduler candidate to ensure network fairness.

The difference between GADaM's prototype and round-robin is clear to be seen in Figure 4.17. GADaM's result achieves a significantly lower average download completion time. Besides, in general, GADaM's result seems more stable than round-robin except in the 8 MB case. However, in the 16 MB case, although it achieved remarkable improvement in terms of variance and average, GADaM also possessed some significant outliers.

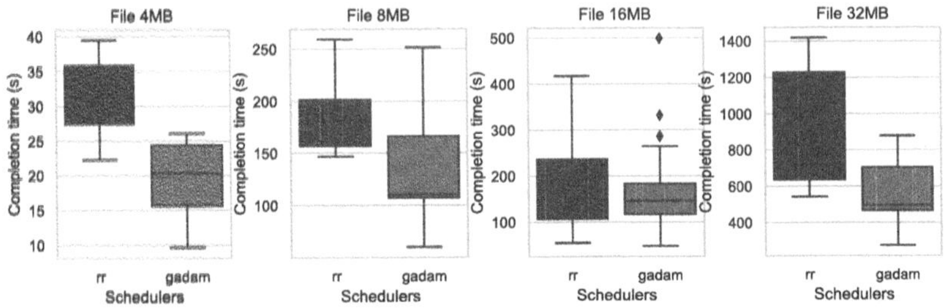

Figure 4.17. *Round-robin versus GADaM scheduler performances. For a color version of this figure, see www.iste.co.uk/chu/softwaredefined.zip*

We now move the discussion to our second scheduler candidate, which is minRTT, the chart of which is plotted in Figure 4.18. The equivalent trend of GADaM's improvement, which happened with round-robin, appears to be equivalent with minRTT: considering lower average download completion time, lower variance and some noticeable outliers. However, in this case, minRTT also possessed these outliers.

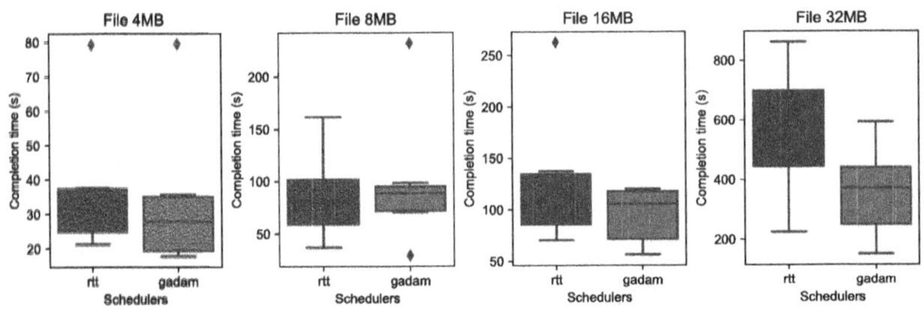

Figure 4.18. *MinRTT versus GADaM schedulers performances. For a color version of this figure, see www.iste.co.uk/chu/softwaredefined.zip*

The third candidate will be BLEST versus GADaM in Figure 4.19. The trend in this run no longer seems to be the same: even though it still achieved a significant average download completion time, the result varies dramatically for GADaM. BLEST's performance might not be better in terms of the average but seems much more reliable with significantly lower variance in this particular environment.

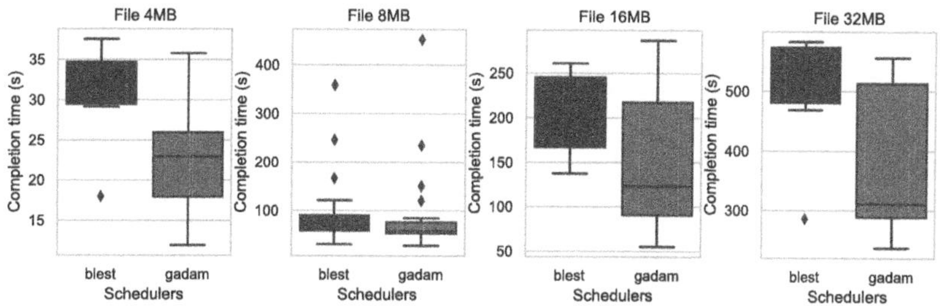

Figure 4.19. *BLEST versus GADaM scheduler performances. For a color version of this figure, see www.iste.co.uk/chu/softwaredefined.zip*

In the comparison between ECF and GADaM, Figure 4.20 shows that GADaM achieves greater overall stability with fewer outliers while maintaining a download completion time comparable to ECF.

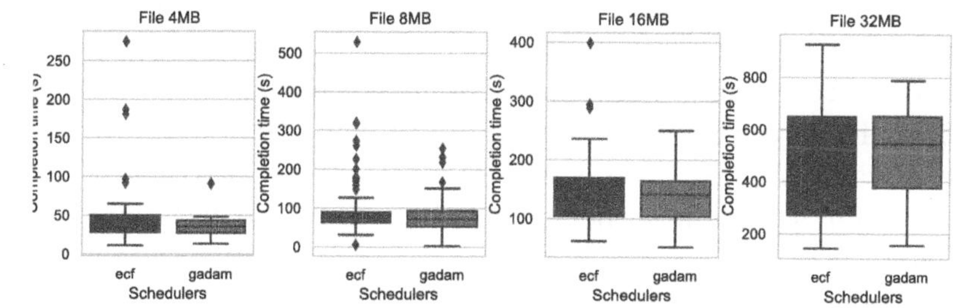

Figure 4.20. *ECF versus GADaM scheduler performances. For a color version of this figure, see www.iste.co.uk/chu/softwaredefined.zip*

The last and also the most important comparison is between GADaM and Peekaboo in Figure 4.21. GADaM might have some advantages in the 4 MB case; however, in other file sizes, GADaM's performance is mostly overtaken by Peekaboo with significant download completion time and lower variance.

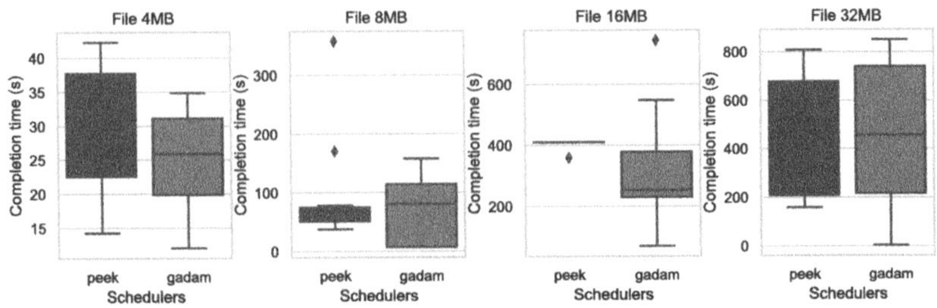

Figure 4.21. *Peekaboo versus GADaM scheduler performances. For a color version of this figure, see www.iste.co.uk/chu/softwaredefined.zip*

As we have just witnessed, GADaM's approach increases the performance stability of each scheduler candidate except for the state-of-the-art comparison, that is, Peekaboo. However, as a reminder, the experimental environment for this analysis is the static in-house environment, which is logical since the proposed improvement of GADaM's approach is for a more unstable, fluctuating, dynamic environment. Nevertheless, as seen in the variation mentioned in Table 4.2, even in this static environment, there is still some variance in network conditions regarding the fixed position of the hardware. However, an argument might indicate that GADaM might always suggest the state of the art to deal with the scheduler selection to perform better than the previous scheduler.

4.5.5. *Actual run*

This section is about the experiment process we conducted to achieve our final result after the trial run's insights.

The only difference between our actual experiment versus our trial run is the client number within the client cluster and the server number within the server cluster. The experiment environment, which includes the mobile executing platform, and the cloud server platform should be fair between each scheduler candidate so distinguishing between each scheduler hardware is necessary. For the network infrastructure, since we are using the same operating environment for each kind of physical interface, the coverage signal is identical and equivalent between each mobile phone.

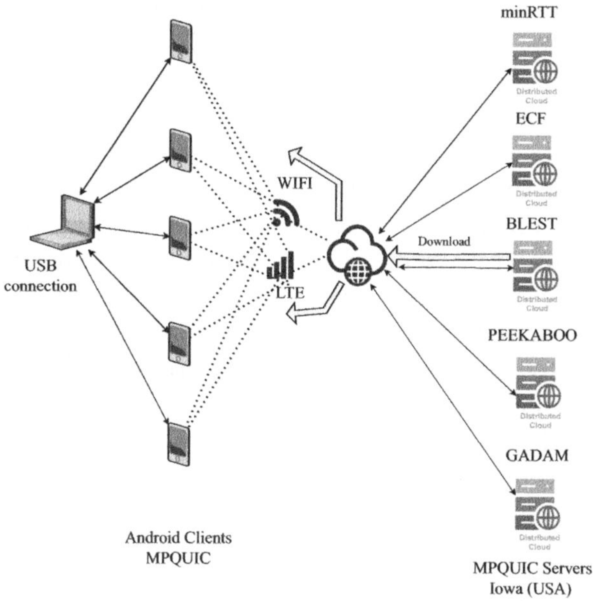

Figure 4.22. *Experimental setup. For a color version of this figure, see www.iste.co.uk/chu/softwaredefined.zip*

After describing our experiment setup, the intensive experiment's analysis, which was conducted in multiple mobile urban scenarios, will be discussed. Due to poor performance, round-robin is excluded from this experiment in order to speed up the experiment process. The following section will explain the in-depth analysis of our collected results.

In this section, the intensive experiment's analysis, which was conducted in multiple mobile urban scenarios, will be discussed. Based on the obtained result, a in-depth analysis is presented and hence proves the effectiveness of the proposed theories.

Motivated by the initial promising result in the trial run, which is significantly improved compared to the predefined scheduler approach, except for the state-of-the-art Peekaboo, we proceed to the next phase of the experiment, which is to experiment with all the candidates at the same time. However, due to poor performance, round-robin is excluded from this experiment to speed up the process. The experiment will be conducted using the same framework but with an increase in the mobile phones and the cloud servers according to the number of schedulers. We decided to conduct our experiment more frequently to have more objective observations. In addition, we persisted in conducting the enhanced experiment in the static environment to better understand the static environment, which could be considered as the baseline. To enable mobility in the second actual run, we used two mobile urban scenarios: in-vehicle and metro lines within the Paris region.

4.5.6. *In-house static environment*

The first result for the static environment is shown in Figure 4.23. While minRTT possessed the highest download completion time with huge variance, BLEST and ECF maintained their positions, respectively. Peekaboo still achieved a remarkable performance compared to its previous one. However, there are some positive improvements for GADaM's performance in this environment except in the 8 MB file size: the best average download completion time and a remarkably lower variance compared to the state-of-the-art Peekaboo in 4 MB and 16 MB.

A remark that should be made for this result is, even though expected not to have any significant improvement in terms of proposed metrics, GADaM's approach could however be considered to outperform the predefined state-of-the-art scheduler suggestion. Further interesting insights could be seen in our later experiment results.

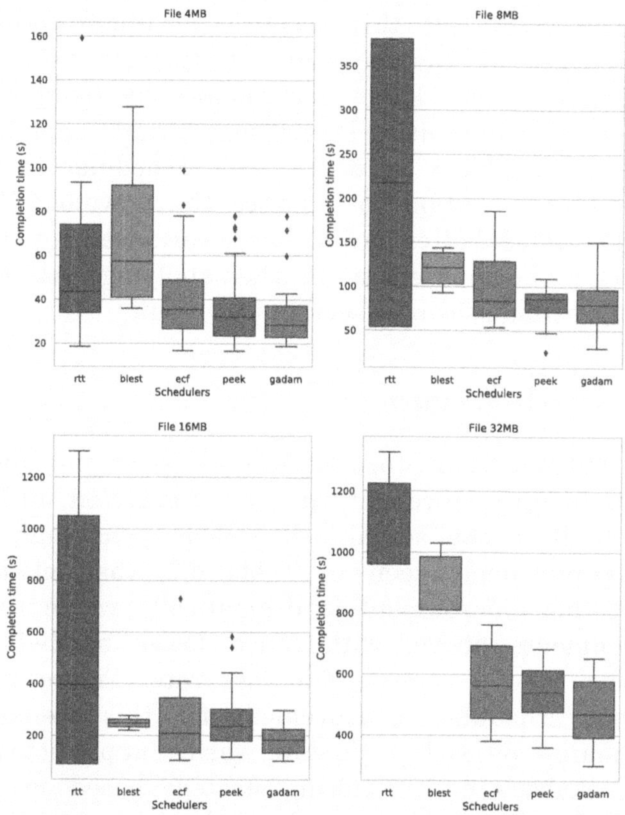

Figure 4.23. *Static scenario scheduler performances. For a color version of this figure, see www.iste.co.uk/chu/softwaredefined.zip*

4.5.7. *Metro line environment*

In this environment, the mobility of the metro could be considered the most crucial factor affecting the performance measurement. The network condition in this environment is much improved compared to the static in-house environment, as in Table 4.2. The asymmetric conditions seem not to be the case in this environment. However, because the duration for our network measurement is only the first 10 s, the circumstances may vary depending on the actual situation. The result for this scenario is illustrated in Figure 4.24. Even in this environment, the deterministic approach achieved remarkable performance by fixing the state-of-the-art scheduler using Peekaboo. GADaM's approach has proved its capability to utilize each scheduler in its suitable network condition to enhance the download file performance. A much better metric, lower variance with only a few outliers, is the apparent trend when applying GADaM's scheduler suggestion as can be seen in this result.

4.5.8. *In-vehicle environment*

Our last but not least proposed scenario is an in-vehicle environment, mainly since our run was conducted during rush hour to have the actual urban density as high as possible. As previously seen in Table 4.2, the network condition of this environment is dramatically worse than both environments above, with much lower bandwidth, higher latency and a considerable loss rate. Figure 4.25 describes our scheduler measurement for this scenario. The performance of each candidate varies unpredictably. However, Peekaboo's path-exploring strategy based on the multi-armed bandit problem might help it become the only acceptable candidate in this environment. GADaM's performance might achieve the best in the smallest file size; however, the larger the file size, the lower its performance is, compared to the fixed state-of-the-art approach. In addition to this, in our last scenario with the largest file size 32 MB, we witnessed the experiment process most of the time exceed our time-out threshold, hence the experiment results for minRTT, BLEST and ECF were forced to shut down. Peekaboo and GADaM were the only survivors in this particular case. Our first thought was that, the reason for GADaM's performance might come from too poor network conditions. However, by cross-checking our experiment network conditions with our previous work, the dataset we used to train for GADaM's prototype model has no relation to these network characteristics. Put another way, GADaM's prototype knowledge has exact no clue about how to suggest a reasonable scheduler if it is just based on the prior knowledge in its trained environment.

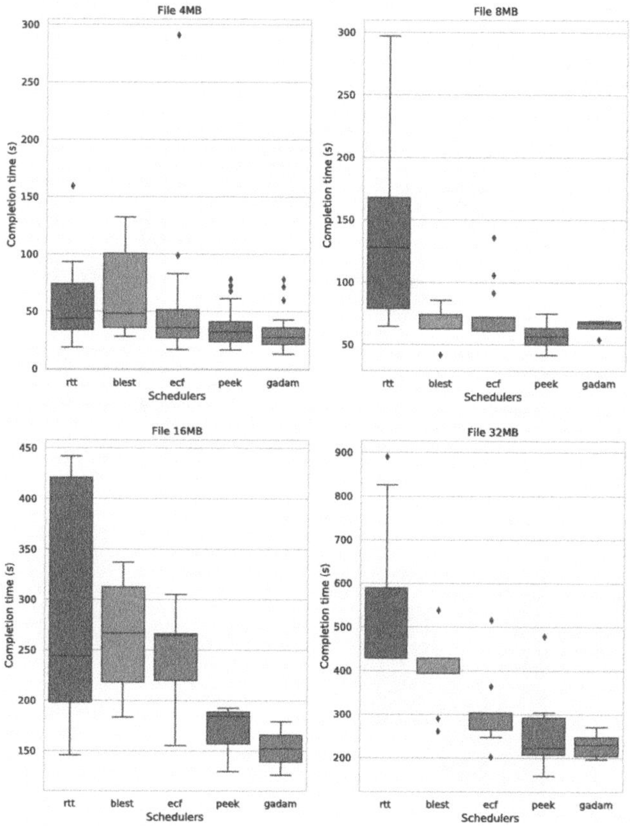

Figure 4.24. *Subway scenario scheduler performances. For a color version of this figure, see www.iste.co.uk/chu/softwaredefined.zip*

After all the analyses, GADaM's approach has proved its improvement capability in its proposed environment, which is mobile, with high dynamicity and frequent fluctuations and reasonable network capacity. The result also confirms the operating capability of the prototype GADaM model in an untrained environment such as high urban density in a small file transmission scenario.

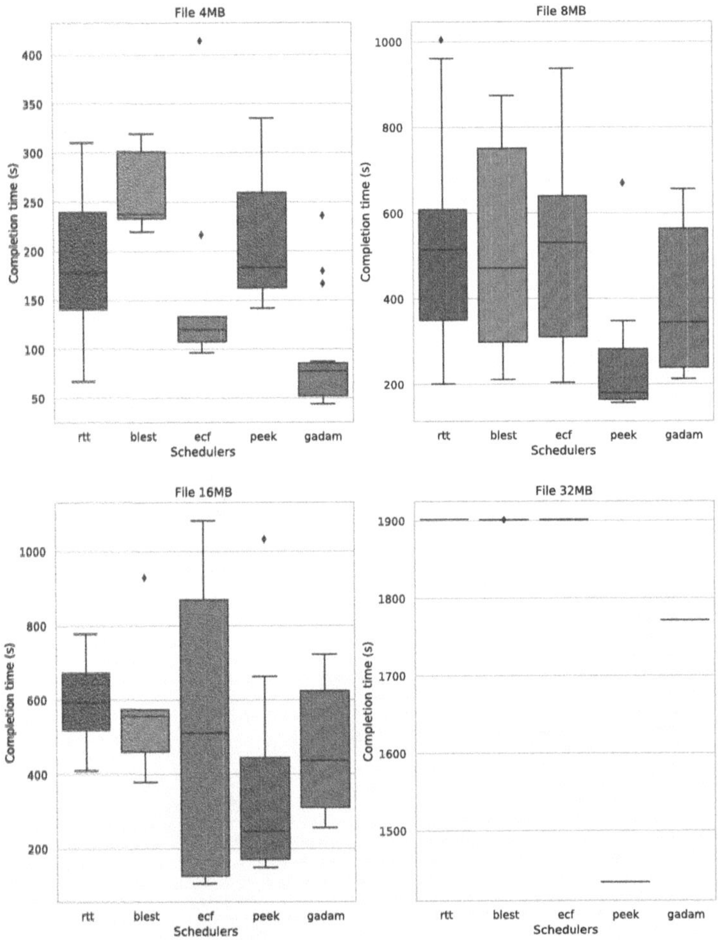

Figure 4.25. *Vehicle scenario scheduler performances. For a color version of this figure, see www.iste.co.uk/chu/softwaredefined.zip*

4.6. Limitations

The advantage of deep learning-based scheduler selectors has been proved in both simulated and practical experiments. The obtained result and in-depth analysis have confirmed the proposed theory in Chu et al. (2022). For the moment, the approach seems heavy compared to the deterministic approach. However, in order to deploy the solution into reality, a lot of considerations need to be seriously taken care of by the operators.

The first and the most important consideration is the mandatory pre-training process. In order to build the dataset, the complexity of the model, as well as the target environment accuracy, hugely depends on the step in each path characteristic parameters, which might dramatically increase the number of required experiments just to collect the performance measurement. Afterward, the dataset consists of many thousands of records, which take an enormous amount of time to conduct within a small laboratory environment, and need to be cleaned and organized. The next important step in this process is finding a deep learning architecture that is capable of classifying the most suitable network condition for each scheduler, which also takes a lot of engineering time. In addition to this, the training process of the model also acquired a huge computation capacity as well as the mandatory time until the model converges. We assumed after all these processes that a fine-tuned model had been conducted.

The second part, which is no less important compared to the first one, is the process to introduce the proposed model into reality. In order to deploy the model in practice, it also depends on the state of the implementation of a multipath transport protocol, deep-learning engine, relevant programming libraries, kernel support by the manufacturer, etc. To deploy GADaM, which is a deep learning model, we thus need the equivalent steps to deploy and an update each time a novel scheduler is invented.

We recap some of the main concerns of the approach as follows:

– complexity of design target dataset;

– time complexity to build the dataset;

– complexity of fine-tuning the model;

– complexity of finding the model;

– complexity of deploying the model;

– complexity of repeating the same process if a novel scheduler is invented.

To resolve the aforementioned concerns, a few approaches have been introduced by the community recently such as Wu et al. (2021a) and Wu et al. (2022). However, we also proposed another approach. To resolve the complexity of designing the target dataset, the time complexity to build the dataset as well as fine-tuning the model, we suggest taking advantage of the online reinforcement learning approach enhanced by the federated learning technique to update our proposed model. The prototype model could be considered as the initial model for further investigation, which helps the training process to get over cold-start problems, but then will adapt itself to each deployed environment.

Furthermore, we only proposed a prototype of GADaM evaluation based on a per packet basis. The effectiveness of GADaM is also required to be evaluated in other scheduler recommendation architectures such as per timeslot or on significant network characteristic changes only.

Our results were collected in multiple environments; however, considering mobile scenarios, plenty of other environments need to be investigated, such as highway automobiles (Chao et al. 2021), aircraft (Luo et al. 2019), in-space transmission (Prochniewicz and Grzymala 2021), unmanned aerial vehicles (Li et al. 2019a) and high-speed railways (Jaffry et al. 2021).

Last but not least, an important point to make is the main idea of GADaM is selecting the most suitable scheduler, which means GADaM's performance is hugely dependent on its candidate's performance. It is fantasy to expect GADaM to have a remarkable result when all the candidate schedulers' performances are poor within the same scenario. These remarks also emphasize that GADaM's improvement might be much more significant in frequently changing network conditions. For sure, further investigation into scheduler candidates, especially in learning branch, is required to achieve a better overall scheduling performance. These ideas should be considered as potential future work.

4.7. Conclusion

In this chapter, we identified one missing piece of the current multipath research direction, proposed a paradigm to tackle this issue and steer a cutting-edge transport protocol to assist the best network performance for healthcare emergency communications.

The GADaM prototype is implemented using MP-QUIC's implementation and used to evaluate the classification performance of the proposed architecture in both simulated and practical environments. In a simulated environment, the proposed deep learning architecture selected the highest performance scheduler based on the current network characteristics with more than 95% accuracy on the training set and 91% accuracy on the testing set. Meanwhile, we presented an analytical and practical experiment in the mobile environment of GADaM's prototype implementation. In order to conduct our experiment, an extensible modular scheduler evaluating framework was also designed and implemented. In practical evaluation, GADaM's improvement was more significant in long-term transmission in static and mobile experiment environments. Even in undesired network conditions in which GADaM has no prior knowledge, GADaM also provided the best performance in short-term transmission compared to the state of the art. The result confirmed the proposed theories in our previous work and proved its advantages compared to the deterministic scheduler selection approach. However, as mentioned before, to have a more solid understanding of the suggested proposal, more investigation must be considered for our future work.

The final chapter will recap the thesis overview and our navigations to develop our research in the H-IoT domain further.

Conclusions and Perspectives

We are all now connected
by the Internet, like
neurons in a giant brain

Stephen Hawking

C.1. Summary of contributions

With the vast development of the Internet of Things and global health concerns, H-IoT will become more and more crucial and impactful to modern society. This fact motivated us to study the domain's state-of-the-art and look for the missing pieces in the community's vision. While conducting our research, we also considered the strictest scenarios in the H-IoT domain, healthcare emergencies, to preserve the comprehensiveness of our solution. However, our proposed solutions are feasible to apply to the general IoT context or even less challenging scenarios and other domains. Thanks to the cutting-edge emergence of SDN and edge computing, various opportunities and applications in this domain have been explored for the prospective evolution. Despite significant research, the gap between the domain's current landscape and its fulfilled prospect remains prominent. Striving to fill this gap, we use an incremental approach to propose in this thesis three coherent contributions, which integrate three levels from the infrastructure to the terminal node.

The first level concerns the network infrastructure and manages all types of traffic in a mobile and constrained environment. We have proposed a unified functional model built around software-defined infrastructure (SDI) "blocks", enabling us to define a logical architecture above the physical infrastructure. This unified overview allows us to overcome the limitations imposed by the heterogeneity of the equipment used in any H-IoT infrastructure and, more generally, any IoT infrastructure. Several use cases were tested by simulation, validating the relevance of the proposed SDI approach. The proposed approach can also be applied to several other domains, given its genericity. We then moved down a notch in the proposed communication layer to address dynamicity in network infrastructure, particularly in service migration for a mobile object. Such an object could be represented by an emergency vehicle, such as a SAMU (French emergency services) or other, on the move for pre-hospital patient care. We have developed a technique that takes advantage of the proposed SDI network overview in the first level to cut the traditional migration time in half, representing an improvement in terms of overall latency. The third level consisted of implementing a new embedded solution within the mobile object to further optimize end-to-end latency. In this analysis, we proposed a new approach based on machine learning called generic adaptive deep-learning-based multipath scheduler selector (GADaM), which selects the most optimal multipath scheduler based on current network characteristics to manage multiple network interfaces for sending vital healthcare data. Such an adaptive selection approach improves the connection of the mobile object (here, the emergency services vehicle) under dynamic conditions without external assistance. In the simulation environment, the results obtained from various practical mobile scenarios also confirmed the effectiveness of the proposed approach in highly dynamic and constrained environments.

C.2. Perspectives and future work

Taking into account the presented SDI paradigm and the implemented functional blocks, the studied use cases mainly focused on supporting healthcare transport with four vehicle scenarios in a road environment, partially enhanced by UAV deployment. Other operational environments (air, space, land, underwater, etc.) and use cases in various domains are promising directions for extending our approach. Considering the state of the art, the SDI paradigm also requires huge initiatives from the community to standardize related technical aspects and implementation perspectives. The proposed solution offers various opportunities for optimizing the communication layer, thanks to SDI's unlimited orchestration capability and accessibility.

Furthermore, several perspectives correlated to the proposed SDI paradigm usage in context are available. For example, its application in a critical H-IoT situation to maintain patient accessibility by optimizing the latency of on-board applications in emergency vehicles. More precisely, the proposed SDI controller could activate its unified functional model and then combine it with connection migration to the optimal edge computing servers to optimize the ambulance's network performance, which is feasible. Notably, an AI-based solution and a practical evaluation of the "connection migration" paradigm in various mobile network environments are highly recommended to study its impact further. Regarding the data privacy concerns, the SDI controller could play the role of a trustworthy authenticator to verify participants before granting them authorized access to healthcare data and withdraw the permission when it is not necessary. Besides, the SDI-enabled connection migration ensures the communication channel's integrity between authorized parties. Hence, the combination of SDI and connection migration is also capable of isolating healthcare communication from other domains' network activities. Further study in this direction is considered since the domain's data nature is sensitive. Nevertheless, each on-board network device will fully exploit its multiple network interfaces as a stand-alone solution or in a controlled manner to improve overall system performance.

References

3GPP (2020). 23.501: System architecture for the 5g system v16.4. Specification 23.501, 3GPP, Sophia Antipolis [Online]. Available at: https://portal.3gpp.org/desktopmodules/Specifications/SpecificationDetails.aspx?specificationId=3144.

Abdel-Basset, M., Manogaran, G., Gamal, A., Chang, V. (2020). A novel intelligent medical decision support model based on soft computing and IoT. *IEEE Internet of Things Journal*, 7(5), 4160–4170.

Abdellatif, A.A., Mohamed, A., Chiasserini, C.F., Tlili, M., Erbad, A. (2020). Edge computing for smart health: Context-aware approaches, opportunities, and challenges. *IEEE Network*, 33(3), 196–203. doi: 10.1109/MNET.2019.1800083.

Abkenar, F.S., Iranmanesh, S., Bouguettaya, A., Raad, R., Jamalipour, A. (2022). ENERGENT: An energy-efficient UAV-assisted fog-IoT framework for disaster management. *Journal of Communications and Networks*, 24(6), 698–709.

Aggarwal, S., Saha, S.K., Khan, I., Pathak, R., Koutsonikolas, D., Widmer, J. (2022). MuSher: An agile multipath-TCP scheduler for dual-band 802.11ad/ac wireless LANs. *IEEE/ACM Transactions on Networking*, 30(4), 1879–1894.

Ahad, A., Tahir, M., Sheikh, M.A., Ahmed, K.I., Mughees, A., Numani, A. (2020). Technologies trend towards 5G network for smart health-care using IoT: A review. *Sensors*, 20(14), 1–22.

Ahmad, R., Hämäläinen, M., Wazirali, R., Abu-Ain, T. (2023). Digital-care in next generation networks: Requirements and future directions. *Computer Networks*, 224, 109599.

Al-Dhuraibi, Y., Paraiso, F., Djarallah, N., Merle, P. (2018). Elasticity in cloud computing: State of the art and research challenges. *IEEE Transactions on Services Computing*, 11(2), 430–447.

Alharthi, M., Taha, A.E.M., Hassanein, H.S. (2019). An architecture for software defined drone networks. In *IEEE International Conference on Communications*, Shanghai.

Alzahrani, A.O. and Alenazi, M.J. (2023). ML-IDSDN: Machine learning based intrusion detection system for software-defined network. *Concurrency and Computation: Practice and Experience*, 35(1), e7438.

Asadollahi, S., Goswami, B., Sameer, M. (2018). Ryu controller's scalability experiment on software defined networks. In *2018 IEEE International Conference on Current Trends in Advanced Computing, ICCTAC 2018*, Bangalore.

Ashton, K. (2009). That "Internet of Things" thing. *RFID Journal*, June 22 [Online]. Available at: https://www.rfidjournal.com/expert-views/that-internet-of-things-thing/73881/.

Askari, Z., Abouei, J., Jaseemuddin, M., Anpalagan, A. (2021). Energy-efficient and real-time NOMA scheduling in IoMT-based three-tier WBANs. *IEEE Internet of Things Journal*, 8(18), 13975–13990.

Attaran, M. (2023). The impact of 5G on the evolution of intelligent automation and industry digitization. *Journal of Ambient Intelligence and Humanized Computing*, 14, 5977–5993.

Aujla, G.S., Chaudhary, R., Kaur, K., Garg, S., Kumar, N., Ranjan, R. (2019). Safe: SDN-assisted framework for edge-cloud interplay in secure healthcare ecosystem. *IEEE Transactions on Industrial Informatics*, 15(1), 469–480.

Azimi, I., Takalo-Mattila, J., Anzanpour, A., Rahmani, A.M., Soininen, J.P., Liljeberg, P. (2019). Empowering healthcare IoT systems with hierarchical edge-based deep learning. In *2018 IEEE/ACM International Conference on Connected Health: Applications, Systems and Engineering Technologies (CHASE)*, Washington, D.C.

Baktir, A.C., Tunca, C., Ozgovde, A., Salur, G., Ersoy, C. (2018). SDN-based multi-tier computing and communication architecture for pervasive healthcare. *IEEE Access*, 6, 56765–56781.

Bannour, F., Souihi, S., Mellouk, A. (2018). Distributed SDN control: Survey, taxonomy, and challenges. *IEEE Communications Surveys & Tutorials*, 20(1), 333–354.

Bannour, F., Souihi, S., Mellouk, A. (2022). *Software-Defined Networking 2: Extending SDN Control to Large-Scale Networks*. ISTE Ltd, London, and John Wiley & Sons, New York.

Barakabitze, A.A., Ahmad, A., Mijumbi, R., Hines, A. (2020). 5G network slicing using SDN and NFV: A survey of taxonomy, architectures and future challenges. *Computer Networks*, 167, 106984.

Barbarulo, F., Puliafito, C., Virdis, A., Mingozzi, E. (2022a). Enabling application relocation in ETSI MEC: A container-migration approach. In *2022 IEEE 33rd Annual International Symposium on Personal, Indoor and Mobile Radio Communications (PIMRC)*, Kyoto.

Barbarulo, F., Puliafito, C., Virdis, A., Mingozzi, E. (2022b). Extending ETSI MEC towards stateful application relocation based on container migration. In *2022 IEEE 23rd International Symposium on a World of Wireless, Mobile and Multimedia Networks (WoWMoM)*, Belfast.

Beigi-Mohammadi, N., Shtern, M., Litoiu, M. (2020). Adaptive load management of web applications on software defined infrastructure. *IEEE Transactions on Network and Service Management*, 17(1), 488–502.

Benaissa, K., Bitam, S., Mellouk, A. (2021). On-board data management layer: Connected vehicle as data platform. *Electronics*, 10(15), 1810.

Benelmir, R., Bitam, S., Mellouk, A. (2022). Deep learning-based beam pair angle prediction for beyond 5G millimeter-wave vehicular communication. In *GLOBECOM 2022–2022 IEEE Global Communications Conference*, Rio de Janeiro.

Bitam, S., Zeadally, S., Mellouk, A. (2018). Fog computing job scheduling optimization based on bees swarm. *Enterprise Information Systems*, 12(4), 373–397.

Bottou, L. (2003). Stochastic learning. In *Advanced Lectures on Machine Learning*, Bousquet, O., Luxburg, U., Rätsch, G. (eds). Springer, Berlin, Heidelberg.

Bouazizi, A., Zaibi, G., Samet, M., Kachouri, A. (2017). Wireless body area network for e-health applications: Overview. In *2017 International Conference on Smart, Monitored and Controlled Cities (SM2C)*, Sfax.

Campolo, C., Fontes, R.D.R., Molinaro, A., Rothenberg, C.E., Iera, A. (2018). Slicing on the road: Enabling the automotive vertical through 5G network softwarization. *Sensors*, 18(12), 4435.

Chao, L., Wu, C., Yoshinaga, T., Bao, W., Ji, Y. (2021). A brief review of multipath TCP for vehicular networks. *Sensors*, 21(8), 1–34.

Chen, M., Li, W., Hao, Y., Qian, Y., Humar, I. (2018). Edge cognitive computing based smart healthcare system. *Future Generation Computer Systems*, 86, 403–411.

Chen, W., Lin, X., Lee, J., Toskala, A., Sun, S., Chiasserini, C.F., Liu, L. (2023). 5G-advanced toward 6G: Past, present, and future. *IEEE Journal on Selected Areas in Communications*, 41(6), 1592–1619.

Chu, T.-T., Labiod, M.A., Tran, H.-A., Mellouk, A. (2022). Gadam: Generic adaptive deep-learning-based multipath scheduler selector for dynamic heterogeneous environment. In *ICC 2022–IEEE International Conference on Communications*, Seoul.

Chu, T.-T., Labiod, M.A., Augustin, B., Mellouk, A. (2023). Gadam on the road-smart approach to multi-access networks: Analytical and practical evaluation in various urban mobile environments. In *2023 IEEE Wireless Communications and Networking Conference (WCNC)*, Glasgow.

Cicioğu, M. and Çalhan, A. (2019a). Hubsflow: A novel interface protocol for SDN-enabled wbans. *Computer Networks*, 160, 105–117.

Cicioğlu, M. and Çalhan, A. (2019b). SDN-based wireless body area network routing algorithm for healthcare architecture. *ETRI Journal*, 41(4), 452–464.

Conforti, L., Virdis, A., Puliafito, C., Mingozzi, E. (2021). Extending the quic protocol to support live container migration at the edge. In *2021 IEEE 22nd International Symposium on a World of Wireless, Mobile and Multimedia Networks (WoWMoM)*, Pisa.

Dai, X., Spasić, I., Meyer, B., Chapman, S., Andres, F. (2019). Machine learning on mobile: An on-device inference app for skin cancer detection. In *2019 Fourth International Conference on Fog and Mobile Edge Computing (FMEC)*, Rome.

Dai, B., Niu, J., Ren, T., Hu, Z., Atiquzzaman, M. (2022). Towards energy-efficient scheduling of UAV and base station hybrid enabled mobile edge computing. *IEEE Transactions on Vehicular Technology*, 71(1), 915–930.

De Coninck, Q. and Bonaventure, O. (2017). Multipath QUIC: Design and evaluation. In *CoNEXT 2017 – Proceedings of the 2017 13th International Conference on emerging Networking EXperiments and Technologies*. ACM, Incheon.

De Coninck, Q. and Bonaventure, O. (2019). MultipathTester: Comparing MPTCP and MPQUIC in mobile environments. In *TMA 2019 – Proceedings of the 3rd Network Traffic Measurement and Analysis Conference*, Paris.

Devarajan, M., Subramaniyaswamy, V., Vijayakumar, V., Ravi, L. (2019). Fog-assisted personalized healthcare-support system for remote patients with diabetes. *Journal of Ambient Intelligence and Humanized Computing*, 10, 3747–3760.

Diakhate, P.B., Chu, T.-T., Labiod, M.A., Augustin, B., Tran, H.-A., Mellouk, A. (2022). Experimental evaluation of multiple multipath schedulers over various urban mobile environments. In *Proceedings of the Eleventh International Symposium on Information and Communication Technology*. ACM, Hanoi.

Dias, J.A., Rodrigues, J.J., Kumar, N., Korotaev, V., Han, G. (2017). REMA: A REsource MAnagement tool to improve the performance of vehicular delay-tolerant networks. *Vehicular Communications*, 9, 135–143.

Dilley, J., Maggs, B., Parikh, J., Prokop, H., Sitaraman, R., Weihl, B. (2002). Globally distributed content delivery. *IEEE Internet Computing*, 6(5), 50–58.

El Jaouhari, S. and Bouabdallah, A. (2018). Dynamic security management of smart WoT infrastructures using SDN. In *IEEE 88th Vehicular Technology Conference (VTC-Fall)*, Chicago. doi: 10.1109/VTCFall.2018.8690740.

Elayan, H., Shubair, R.M., Kiourti, A. (2017). Wireless sensors for medical applications: Current status and future challenges. In *2017 11th European Conference on Antennas and Propagation (EUCAP)*, Paris.

Eswaran, S. and Honnavalli, P. (2023). Private 5G networks: A survey on enabling technologies, deployment models, use cases and research directions. *Telecommun. Syst.*, 82, 3–26.

Ferlin, S., Alay, O., Mehani, O., Boreli, R. (2016). BLEST: Blocking estimation-based MPTCP scheduler for heterogeneous networks. In *2016 IFIP Networking Conference (IFIP Networking) and Workshops, IFIP Networking 2016*, Vienna.

Fogli, M., Giannelli, C., Stefanelli, C. (2022). Software-defined networking in wireless ad hoc scenarios: Objectives and control architectures. *Journal of Network and Computer Applications*, 203(C), 103387.

da Fonseca, M.H., Kovaleski, F., Picinin, C.T., Pedroso, B., Rubbo, P. (2021). E-Health practices and technologies: A systematic review from 2014 to 2019. *Healthcare*, 9(9), 1192. doi: 10.3390/healthcare9091192.

Fontes, R.R., Afzal, S., Brito, S.H., Santos, M.A., Rothenberg, C.E. (2015). Mininet-WiFi: Emulating software-defined wireless networks. In *Proceedings of the 11th International Conference on Network and Service Management, CNSM 2015*, Barcelona.

Ford, A., Raiciu, C., Handley, M., Bonaventure, O. (2013). Tcp extensions for multipath operation with multiple addresses. *RFC*, 8684, 1–68.

Greco, L., Ritrovato, P., Xhafa, F. (2019). An edge-stream computing infrastructure for real-time analysis of wearable sensors data. *Future Generation Computer Systems*, 93, 515–528.

Halman, L.M. and Alenazi, M.J. (2023). MCAD: A machine learning based cyberattacks detector in software-defined networking (SDN) for healthcare systems. *IEEE Access*, 11, 37052–37067.

Handigol, N., Heller, B., Jeyakumar, V., Lantz, B., McKeown, N. (2012). Reproducible network experiments using container-based emulation. In *CoNEXT 2012 – Proceedings of the 2012 ACM Conference on Emerging Networking Experiments and Technologies*, Nice.

Hasan, K., Ahmed, K., Biswas, K., Islam, M.S., Kayes, A.S., Islam, S.M.R. (2020a). Control plane optimisation for an SDN-based WBAN framework to support healthcare applications. *Sensors*, 20(15), 1–19.

Hasan, K., Ahmed, K., Biswas, K., Islam, M.S., Sianaki, O.A. (2020b). Software-defined application-specific traffic management for wireless body area networks. *Future Generation Computer Systems*, 107, 274–285.

Hayajneh, T., Griggs, K., Imran, M., Mohd, B.J. (2019). Secure and efficient data delivery for fog-assisted wireless body area networks. *Peer-to-Peer Networking and Applications*, 12, 1289–1307.

Hazra, A., Rana, P., Adhikari, M., Amgoth, T. (2023). Fog computing for next-generation internet of things: Fundamental, state-of-the-art and research challenges. *Computer Science Review*, 48(C), 100549.

Hegde, C., Jiang, Z., Suresha, P.B., Zelko, J., Seyedi, S., Smith, M.A., Wright, D.W., Kamaleswaran, R., Reyna, M.A., Clifford, G.D. (2020). Autotriage – An open source edge computing raspberry pi-based clinical screening system. *medRxiv*, 216054925 [Online]. Available at: https://www.medrxiv.org/content/early/2020/04/30/2020.04.09.20059840.

Hemminger, S. (2005). Network emulation with NetEm. In *Proceedings of the 6th Australian National Linux Conference (LCA 2005)*, Canberra.

Hu, J., Wu, K., Liang, W. (2019). An IPv6-based framework for fog-assisted healthcare monitoring. *Advances in Mechanical Engineering*, 11(1). doi: 10.1177/1687814018819515.

Huang, G., Luo, C., Wu, K., Ma, Y., Zhang, Y., Liu, X. (2019). Software-defined infrastructure for decentralized data lifecycle governance: Principled design and open challenges. In *2019 IEEE 39th International Conference on Distributed Computing Systems (ICDCS)*, Dallas.

Hui, P., Chaintreau, A., Scott, J., Gass, R., Crowcroft, J., Diot, C. (2005). Pocket switched networks and human mobility in conference environments. In *Proceedings of the 2005 ACM SIGCOMM Workshop on Delay-Tolerant Networking, WDTN '05, Association for Computing Machinery*. ACM, New York.

Islam, M.M., Khan, M.T.R., Saad, M.M., Kim, D. (2021). Software-defined vehicular network (SDVN): A survey on architecture and routing. *Journal of Systems Architecture*, 114(C), 101961.

Iyengar, J.R. and Thomson, M. (2021). RFC 9000: QUIC: A UDP-based multiplexed and secure transport. Standard, RFC Editor, Marina del Rey [Online]. Available at: https://www.rfc-editor.org/info/rfc9000.

Iyengar, J.R., Amer, P.D., Stewart, R. (2006). Concurrent multipath transfer using SCTP multihoming over independent end-to-end paths. *IEEE/ACM Transactions on Networking*, 14(5), 951–964.

Ja'afreh, M.A., Adhami, H., Alchalabi, A.E., Hoda, M., Saddik, A.E. (2022). Toward integrating software-defined networks with the Internet of Things: A review. *Cluster Computing*, 25, 1619–1636.

Jaffry, S., Hussain, R., Gui, X., Hasan, S.F. (2021). A comprehensive survey on moving networks. *IEEE Communications Surveys and Tutorials*, 23(1), 110–136.

Jameel, F., Khan, W.U., Jamshed, M.A., Pervaiz, H., Abbasi, Q., Jantti, R. (2020). Reinforcement learning for scalable and reliable power allocation in SDN-based backscatter heterogeneous network. In *IEEE INFOCOM 2020 – IEEE Conference on Computer Communications Workshops (INFOCOM WKSHPS)*, Toronto.

Jangra, P. and Gupta, M. (2018). A design of real-time multilayered smart healthcare monitoring framework using IoT. In *2018 International Conference on Intelligent and Advanced System (ICIAS)*, Kuala Lumpur.

Kafetzis, D., Vassilaras, S., Vardoulias, G., Koutsopoulos, I. (2022). Software-defined networking meets software-defined radio in mobile ad hoc networks: State of the art and future directions. *IEEE Access*, 10, 9989–10014.

Kang, J.-M., Lin, T., Bannazadeh, H., Leon-Garcia, A. (2014). Software-defined infrastructure and the SAVI testbed. In *Testbeds and Research Infrastructure: Development of Networks and Communities: 9th International ICST Conference, TridentCom 2014*. Springer, Guangzhou.

Khan, W.Z., Ahmed, E., Hakak, S., Yaqoob, I., Ahmed, A. (2019). Edge computing: A survey. *Future Generation Computer Systems*, 97, 219–235.

Khan, A., Gupta, S., Gupta, S.K. (2022). Emerging UAV technology for disaster detection, mitigation, response, and preparedness. *Journal of Field Robotics*, 39(6), 905–955.

Kharel, J., Reda, H.T., Shin, S.Y. (2019). Fog computing-based smart health monitoring system deploying LoRa wireless communication. *IETE Technical Review*, 36(1), 69–82.

Khorsandroo, S., Sánchez, A.G., Tosun, A.S., Arco, J.M., Doriguzzi-Corin, R. (2021). Hybrid SDN evolution: A comprehensive survey of the state-of-the-art. *Computer Networks*, 192, 107981.

Kumari, P. and Kaur, P. (2021). A survey of fault tolerance in cloud computing. *Journal of King Saud University-Computer and Information Sciences*, 33(10), 1159–1176.

Lakhan, A., Mohammed, M.A., Obaid, O.I., Chakraborty, C., Abdulkareem, K.H., Kadry, S. (2022). Efficient deep-reinforcement learning aware resource allocation in SDN-enabled fog paradigm. *Automated Software Engineering*, 29(1), 1–25.

Lebre, M.-A., Le Mouël, F., Ménard, E. (2020). Efficient vehicular crowdsourcing models in VANET for disaster management. In *2020 IEEE 91st Vehicular Technology Conference (VTC2020-Spring)*, Antwerp.

Lee, S. and Yoo, J. (2022). Reinforcement learning based multipath quic scheduler for multimedia streaming. *Sensors*, 22(17), 6333.

Lee, H., Flinn, J., Tonshal, B. (2018). Raven: Improving interactive latency for the connected car. In *Proceedings of the 24th Annual International Conference on Mobile Computing and Networking, MobiCom '18*. ACM, New York. doi: 10.1145/3241539.3241571.

Li, D. (2019). 5G and intelligence medicine–How the next generation of wireless technology will reconstruct healthcare? *Precision Clinical Medicine*, 2(4), 205–208.

Li, B., Fei, Z., Zhang, Y. (2019a). UAV communications for 5G and beyond: Recent advances and future trends. *IEEE Internet of Things Journal*, 6(2), 2241–2263.

Li, P., Xu, C., Jin, H., Hu, C., Luo, Y., Cao, Y., Mathew, J., Ma, Y. (2019b). Chainsdi: A software-defined infrastructure for regulation-compliant home-based healthcare services secured by blockchains. *IEEE Systems Journal*, 14(2), 2042–2053.

Li, J., Cai, J., Khan, F., Rehman, A.U., Balasubramaniam, V., Sun, J., Venu, P. (2020). A secured framework for SDN-based edge computing in IoT-enabled healthcare system. *IEEE Access*, 8, 135479–135490.

Lim, Y.S., Nahum, E.M., Towsley, D., Gibbens, R.J. (2017). ECF: An MPTCP path scheduler to manage heterogeneous paths. *Performance Evaluation Review*, 45(1), 33–34.

Lin, T., Kang, J.M., Bannazadeh, H., Leon-Garcia, A. (2014). Enabling SDN applications on software-defined infrastructure. In *IEEE/IFIP NOMS 2014 – IEEE/IFIP Network Operations and Management Symposium: Management in a Software Defined World*, Krakow.

Liu, C., Cao, Y., Luo, Y., Chen, G., Vokkarane, V., Yunsheng, M., Chen, S., Hou, P. (2018a). A new deep learning-based food recognition system for dietary assessment on an edge computing service infrastructure. *IEEE Transactions on Services Computing*, 11(2), 249–261.

Liu, X., Jia, M., Zhang, X., Lu, W. (2018b). A novel multichannel Internet of Things based on dynamic spectrum sharing in 5G communication. *IEEE Internet of Things Journal*, 6(4), 5962–5970.

Liu, X., Faes, L., Kale, A.U., Wagner, S.K., Fu, D.J., Bruynseels, A., Mahendiran, T., Moraes, G., Shamdas, M., Kern, C. et al. (2019). A comparison of deep learning performance against health-care professionals in detecting diseases from medical imaging: A systematic review and meta-analysis. *The Lancet Digital Health*, 1(6), e271–e297.

Liu, Q., Han, T., Xie, J.L., Kim, B. (2022). Real-time dynamic map with crowdsourcing vehicles in edge computing. *IEEE Transactions on Intelligent Vehicles*, 8(4), 2810–2820.

Luo, Q., Wang, J., Liu, S. (2019). AeroMRP: A multipath reliable transport protocol for aeronautical ad hoc networks. *IEEE Internet of Things Journal*, 6(2), 3399–3410.

Ma, L., Yi, S., Carter, N., Li, Q. (2018). Efficient live migration of edge services leveraging container layered storage. *IEEE Transactions on Mobile Computing*, 18(9), 2020–2033.

Machen, A., Wang, S., Leung, K.K., Ko, B.J., Salonidis, T. (2017). Live service migration in mobile edge clouds. *IEEE Wireless Communications*, 25(1), 140–147.

Mangiante, S., Klas, G., Navon, A., GuanHua, Z., Ran, J., Silva, M.D. (2017). VR is on the edge: How to deliver 360° videos in mobile networks. In *Proceedings of the Workshop on Virtual Reality and Augmented Reality Network*. ACM, New York.

Manoj Kumar, G., Danti, R., Amit, O., Guru Raghavendra, R., Kiran, B., Sanjay, H. (2022). Performance evaluation of HPC application in containerized and virtualized environment. In *Emerging Research in Computing, Information, Communication and Applications: ERCICA 2020*. Springer, Singapore.

McKeown, N. (2014). Software-defined networks and the maturing of the internet. IET, London [Online]. Available at: https://digital-library.theiet.org/do/10.1049/19509/abs/ [Accessed March 19 2025].

Mekki, T., Jabri, I., Rachedi, A., Chaari, L. (2021). Software-defined networking in vehicular networks: A survey. *Transactions on Emerging Telecommunications Technologies*, 33(10), 1–29.

Mellouk, A., Hoceini, S., Zeadally, S. (2013). A state-dependent time evolving multi-constraint routing algorithm. *ACM Transactions on Autonomous and Adaptive Systems (TAAS)*, 8(1), 1–21.

Meng, W., Choo, K.K.R., Furnell, S., Vasilakos, A.V., Probst, C.W. (2018). Towards Bayesian-based trust management for insider attacks in healthcare software-defined networks. *IEEE Transactions on Network and Service Management*, 15(2), 761–773.

Misra, S., Saha, R., Ahmed, N. (2020). Health-flow: Criticality-aware flow control for SDN-based healthcare IoT. In *GLOBECOM 2020–2020 IEEE Global Communications Conference*, Taipei.

Misra, S., Pal, S., Ahmed, N., Mukherjee, A. (2023). SDN-controlled resource-tailored analytics for healthcare IoT system. *IEEE Systems Journal*, 17, 1777–1784.

Morgado, A., Huq, K.M.S., Mumtaz, S., Rodriguez, J. (2018). A survey of 5G technologies: Regulatory, standardization and industrial perspectives. *Digital Communications and Networks*, 4(2), 87–97.

Mozaffari, M., Saad, W., Bennis, M., Nam, Y.H., Debbah, M. (2019). A tutorial on UAVs for wireless networks: Applications, challenges, and open problems. *IEEE Communications Surveys and Tutorials*, 21(3), 2334–2360.

Muhammad, G., Alhamid, M.F., Alsulaiman, M., Gupta, B. (2018a). Edge computing with cloud for voice disorder assessment and treatment. *IEEE Communications Magazine*, 56(4), 60–65.

Muhammad, T., Mehmood, R., Albeshri, A., Katib, I. (2018b). Ubehealth: A personalized ubiquitous cloud and edge-enabled networked healthcare system for smart cities. *IEEE Access*, 6, 32258–32285.

Munawar, H.S., Hammad, A.W.A., Waller, S.T., Thaheem, M.J., Shrestha, A. (2021). An integrated approach for post-disaster flood management via the use of cutting-edge technologies and UAVs: A review. *Sustainability*, 13(14), 7925.

Mushtaq, M.S., Fowler, S., Augustin, B., Mellouk, A. (2016). QoE in 5G cloud networks using multimedia services. In *2016 IEEE Wireless Communications and Networking Conference*, Doha.

Paasch, C., Ferlin, S., Alay, O., Bonaventure, O. (2014). Experimental evaluation of multipath TCP schedulers. In *CSWS 2014 – Proceedings of the ACM SIGCOMM 2014 Workshop on Capacity Sharing Workshop*. ACM, Chicago.

Pace, P., Aloi, G., Gravina, R., Caliciuri, G., Fortino, G., Member, S., Liotta, A. (2018). An edge-based architecture to support efficient applications for healthcare industry 4.0. *IEEE Transactions on Industrial Informatics*, 15(1), 481–489.

Park, S. and Yoo, Y. (2017). Network intelligence based on network state information for connected vehicles utilizing fog computing. *Mobile Information Systems*, 2017, 1–9. doi: 10.1155/2017/7479267.

Parsaeefard, S., Tabrizian, I., Leon-Garcia, A. (2019). Artificial intelligence as a service (AI-aaS) on software-defined infrastructure. In *2019 IEEE Conference on Standards for Communications and Networking (CSCN)*, Granada.

Paszke, A., Gross, S., Massa, F., Lerer, A., Bradbury, J., Chanan, G., Killeen, T., Lin, Z., Gimelshein, N., Antiga, L. et al. (2019). PyTorch: An imperative style, high-performance deep learning library. *Advances in Neural Information Processing Systems*, 32(NeurIPS). doi: 10.48550/arXiv.1912.01703.

Pedregosa, F., Grisel, O., Weiss, R., Passos, A., Brucher, M., Varoquax, G., Gramfort, A., Michel, V., Thirion, B., Grisel, O. et al. (2011). Scikit-learn: Machine learning in Python. *Journal of Machine Learning Research*, 12, 2825–2830.

Pham, M., Mengistu, Y., Do, H., Sheng, W. (2018). Delivering home healthcare through a cloud-based smart home environment (CoSHE). *Future Generation Computer Systems*, 81, 129–140.

Popovski, P., Trillingsgaard, K.F., Simeone, O., Durisi, G. (2018). 5G wireless network slicing for eMBB, URLLC, and mMTC: A communication-theoretic view. *IEEE Access*, 6, 55765–55779.

Preveze, B., Alkhayyat, A., Abedi, F., Jawad, A.M., Abosinnee, A.S. (2022). SDN-driven internet of health things: A novel adaptive switching technique for hospital healthcare monitoring system. *Wireless Communications and Mobile Computing*, 2022(1), 3150756.

Priyadarshini, R., Barik, R.K., Dubey, H. (2018). Deepfog: Fog computing-based deep neural architecture for prediction of stress types, diabetes and hypertension attacks. *Computation*, 6(4) [Online]. Available at: https://www.mdpi.com/2079-3197/6/4/62.

Prochniewicz, D. and Grzymala, M. (2021). Analysis of the impact of multipath on Galileo system measurements. *Remote Sensing*, 13(12), 2295.

Qadri, Y.A., Nauman, A., Zikria, Y.B., Vasilakos, A.V., Kim, S.W. (2020). The future of healthcare Internet of Things: A survey of emerging technologies. *IEEE Communications Surveys & Tutorials*, 22(2), 1121–1167.

Queralta, J.P., Gia, T.N., Tenhunen, H., Westerlund, T. (2019). Edge-AI in LoRa-based health monitoring: Fall detection system with fog computing and LSTM recurrent neural networks. In *2019 42nd International Conference on Telecommunications and Signal Processing (TSP)*, Budapest.

Radoglou-Grammatikis, P., Rompolos, K., Sarigiannidis, P., Argyriou, V., Lagkas, T., Sarigiannidis, A., Goudos, S., Wan, S. (2022). Modeling, detecting, and mitigating threats against industrial healthcare systems: A combined software defined networking and reinforcement learning approach. *IEEE Transactions on Industrial Informatics*, 18(3), 2041–2052.

Rahmani, A.M., Gia, T.N., Negash, B., Anzanpour, A., Azimi, I., Jiang, M., Liljeberg, P. (2018). Exploiting smart e-health gateways at the edge of healthcare Internet-of-Things: A fog computing approach. *Future Generation Computer Systems*, 78(Part 2), 641–658.

Rais, R.N.B., Akbar, M.S., Aazam, M. (2018). Fog-supported Internet of Things (IoTs) architecture for remote patient monitoring systems using wireless body area sensor networks. In *2018 IEEE 16th Intl Conf on Dependable, Autonomic and Secure Computing*, Athens [Online]. Available at: https://ieeexplore.ieee.org/document/8511935/.

Ram, S.S., Apduhan, B., Shiratori, N. (2019). A machine learning framework for edge computing to improve prediction accuracy in mobile health monitoring. In *Computational Science and Its Applications – ICCSA 2019*. Springer, Cham.

Ramalho, F., Neto, A., Santos, K., Filho, J.B., Agoulmine, N. (2015). Enhancing ehealth smart applications: A fog-enabled approach. In *2015 17th International Conference on E-health Networking, Application & Services (HealthCom)*, Boston.

Rejeb, A., Rejeb, K., Simske, S., Treiblmaier, H., Zailani, S. (2022). The big picture on the Internet of Things and the smart city: A review of what we know and what we need to know. *Internet of Things*, 19, 100565.

Salameh, A.I. and El Tarhuni, M. (2022). From 5G to 6G – Challenges, technologies, and applications. *Future Internet*, 14(4), 117.

Sarkar, J.L., Cowlessur, S.K., Ramasamy, V., Pati, B., Selvi, T.M., Panigrahi, C.R., Majumder, B., Verma, R.K., Qureshi, N.M.F. (2023). FogCom: SDN-enabled fog node selection for early detection of communicable diseases. *Journal of King Saud University – Computer and Information Sciences*, 35(9), 101432.

Satija, U., Ramkumar, B., Manikandan, S.M. (2017). Real-time signal quality-aware ECG telemetry system for IoT-based health care monitoring. *IEEE Internet of Things Journal*, 4(3), 815–823.

Shi, H., Cui, Y., Wang, X., Hu, Y., Dai, M., Wang, F., Zheng, K. (2018). STMS: Improving MPTCP throughput under heterogeneous networks. In *Proceedings of the 2018 USENIX Conference on Usenix Annual Technical Conference, USENIX ATC '18*. USENIX Association, Boston.

Siddique, A., Hasan, O., Khalid, F., Shafique, M. (2018). ApproxCS: Near-sensor approximate compressed sensing for IoT-healthcare systems. *arXiv:1811.07330* [Online]. Available at: http://arxiv.org/abs/1811.07330.

Siddiqui, A. (2021). The US and Amazon prone to colossal financial losses during a disastrous internet outage. *Digital Information World*, 2 June [Online]. Available at: https://www.digitalinformationworld.com/2021/06/the-cost-of-shutting-down-the-internet-for-an-hour.html.

Silva, M., Teixeira, P., Gomes, C., Dias, D., Luís, M., Sargento, S. (2021). Exploring software-defined networks for seamless handovers in vehicular networks. *Vehicular Communications*, 31, 100372.

Silvera, E., Sharaby, G., Lorenz, D., Shapira, I. (2009). IP mobility to support live migration of virtual machines across subnets. In *Proceedings of SYSTOR 2009: The Israeli Experimental Systems Conference*. ACM, Haifa.

Singh, S. and Prasad, D. (2022). Wireless body area network (WBAN): A review of schemes and protocols. *Materials Today: Proceedings*, 49(Part 8), 3488–3496.

Sinh, D., Le, L.V., Lin, B.S.P., Tung, L.P. (2018). SDN/NFV – A new approach of deploying network infrastructure for IoT. In *2018 27th Wireless and Optical Communication Conference (WOCC)*, Hualien.

Sisinni, E., Bellagente, P., Depari, A., Flammini, A., Gaffurini, M., Pasetti, M., Rinaldi, S., Ferrari, P. (2023). Assessment of time performance of lightweight virtualization for edge computing applications. In *2023 IEEE 19th International Conference on Factory Communication Systems (WFCS)*, Pavia.

Sood, S.K. and Mahajan, I. (2018). A fog-based healthcare framework for chikungunya. *IEEE Internet of Things Journal*, 5(2), 794–801.

Subedi, P., Alsadoon, A., Prasad, P.W., Rehman, S., Giweli, N., Imran, M., Arif, S. (2021). Network slicing: A next generation 5G perspective. *EURASIP Journal on Wireless Communications and Networking*, 2021(1). doi: 10.1186/s13638-021-01983-7.

Sudhamani, C., Roslee, M., Tiang, J.J., Rehman, A.U. (2023). A survey on 5G coverage improvement techniques: Issues and future challenges. *Sensors*, 23(4), 2356.

Sultan, F., Srinivasan, K., Iyer, D., Iftode, L. (2002). Migratory TCP: Connection migration for service continuity in the internet. In *Proceedings 22nd International Conference on Distributed Computing Systems*, Vienna.

Swartz, J. (2014). Q&A with Stephen Hawking. *USA TODAY* [Online]. Available at: https://eu.usatoday.com/story/tech/2014/12/02/stephen-hawking-intel-technology/18027597/.

Talpur, A., Shaikh, F.K., Baloch, N., Felemban, E., Khelil, A., Alam, M.M. (2019). Validation of wired and wireless interconnected body sensor networks. *Sensors (Switzerland)*, 19(17), 3697.

Tan, L., Su, W., Liu, Y., Gao, X., Li, N., Zhang, W. (2020). Proactive connection migration in QUIC. In *MobiQuitous 2020–17th EAI International Conference on Mobile and Ubiquitous Systems: Computing, Networking and Services*. ACM, Darmstadt.

Tchernykh, A., Schwiegelsohn, U., Talbi, E.-G., Babenko, M. (2019). Towards understanding uncertainty in cloud computing with risks of confidentiality, integrity, and availability. *Journal of Computational Science*, 36, 100581.

Tong, V., Souihi, S., Tran, H.A., Mellouk, A. (2022). SDN-based application-aware segment routing for large-scale network. *IEEE Systems Journal*, 16(3), 4401–4410.

Toufga, S., Owezarski, P., Abdellatif, S., Villemur, T. (2018). An SDN hybrid architecture for vehicular networks: Application to Intelligent Transport System. In *9th European Congress on Embedded Real Time Software And Systems (ERTS)*, Toulouse. doi: 10.48550/arXiv.1712.05307.

Tsypkin, Y.Z. (1973). *Foundations of the Theory of Learning Systems*. Academic Press, New York.

Uddin, M.Z. (2019). A wearable sensor-based activity prediction system to facilitate edge computing in smart healthcare system. *Journal of Parallel and Distributed Computing*, 123, 46–53.

Unal, D., Falleni, S., Demirors, E., Enhos, K., Basagni, S., Melodia, T. (2022). A software-defined underwater acoustic networking platform for underwater vehicles. In *ICC 2022-IEEE International Conference on Communications*, Seoul.

Van, T., Tran, H.A., Souihi, S., Mellouk, A. (2018). Network troubleshooting: Survey, taxonomy and challenges. In *2018 International Conference on Smart Communications in Network Technologies (SaCoNeT)*, El Oued.

Vaniukov, S. and Vaniukov, S. (2023). Reducing it infrastructure costs: 5 ways to optimize it infrastructure. *SOFTERMii* [Online]. Available at: https://www.softermii.com/blog/reducing-it-infrastructure-costs.

Vassilakis, V.G., Moscholios, I.D., Alzahrani, B.A., Logothetis, M.D. (2017). Toward information-centric software-defined cellular networks. In *2017 14th International Conference on Telecommunications (ConTEL)*, Zagreb.

Vora, J., Tanwar, S., Tyagi, S., Kumar, N., Rodrigues, J.J.P.C. (2017). FAAL: Fog computing-based patient monitoring system for ambient assisted living. In *2017 IEEE 19th International Conference on e-Health Networking, Applications and Services (Healthcom)*, Dalian.

Wu, F., Wu, T., Yuce, M.R. (2019). An Internet-of-Things (IoT) network system for connected safety and health monitoring applications. *Sensors (Switzerland)*, 19(1), 21.

Wu, H., Alay, O., Brunstrom, A., Ferlin, S., Caso, G. (2020). Peekaboo: Learning-based multipath scheduling for dynamic heterogeneous environments. *IEEE Journal on Selected Areas in Communications*, 38(10), 2295–2310.

Wu, H., Caso, G., Ferlin, S., Alay, O., Brunstrom, A. (2021a). Multipath scheduling for 5G networks: Evaluation and outlook. *IEEE Communications Magazine*, 59(4), 44–50.

Wu, H., Caso, G., Ferlin, S., Alay, Ö., Brunstrom, A. (2021b). Multipath scheduling for 5G networks: Evaluation and outlook. *IEEE Communications Magazine*, 59(4), 44–50.

Wu, H., Alay, O., Brunstrom, A., Caso, G., Ferlin, S. (2022). FALCON: Fast and accurate multipath scheduling using offline and online learning. *arXiv:2201.08969*. doi: 10.48550/arXiv.2201.08969.

Xu, Y., Gui, G., Gacanin, H., Adachi, F. (2021). A survey on resource allocation for 5G heterogeneous networks: Current research, future trends, and challenges. *IEEE Communications Surveys & Tutorials*, 23(2), 668–695.

Xu, F., Zhang, Z., Feng, J., Qin, Z., Xie, Y. (2022). Efficient deployment of multi-UAV assisted mobile edge computing: A cost and energy perspective. *Transactions on Emerging Telecommunications Technologies*, 33(5), e4453.

Yang, M., Wang, S., Calheiros, R. N., Yang, F. (2018). Survey on QoE assessment approach for network service. *IEEE Access*, 6, 48374–48390.

Yu, Y., Calagna, A., Giaccone, P., Chiasserini, C.F. (2023). TCP connection management for stateful container migration at the network edge. In *2023 21st Mediterranean Communication and Computer Networking Conference (MedComNet)*, Ponza Island.

Zeng, H., Cui, L., Tso, F. P., Zhang, Z. (2022). Optimizing multipath QUIC transmission over heterogeneous paths. *Computer Networks*, 215, 109198 [Online]. Available at: https://www.sciencedirect.com/science/article/pii/S1389128622002894.

Zhang, X., Wang, H., Zhao, H. (2018). An SDN framework for UAV backbone network towards knowledge centric networking. In *IEEE Conference on Computer Communications Workshops (INFOCOM WKSHPS)*, Honolulu.

Zhang, H., Li, W., Gao, S., Wang, X., Ye, B. (2019). ReLeS: A neural adaptive multipath scheduler based on deep reinforcement learning. In *IEEE INFOCOM 2019 – IEEE Conference on Computer Communications*, Paris.

Index

Other titles from

ISTE

in

Networks and Telecommunications

2024

HORVÁTH András, TELEK Miklós
Phase Type Distributions: Theory and Application
(Stochastic Models in Computer Science and Telecommunication Networks Set – Volume 2)

2023

RAVALIMINOARIMALALASON Toky Basilide, RANDIMBINDRAINIBE Falimanana
Queues Applied to Telecoms: Courses and Exercises

TONG Van Van, SOUIHI Sami, TRAN Hai-Anh, MELLOUK Abdelhamid
Troubleshooting for Network Operators: The Road to a New Paradigm with Encrypted Traffic
(New Generation Networks Set – Volume 3)

2022

BANNOUR Fetia, SOUIHI Sami, MELLOUK Abdelhamid
Software-Defined Networking: Extending SDN Control to Large-Scale Networks
(New Generation Networks Set – Volume 2)

BENAROUS Leila, BITAM Salim, MELLOUK Abdelhamid
Security in Vehicular Networks: Focus on Location and Identity Privacy
(New Generation Networks Set – Volume 1)

2021

LAUNAY Frédéric
NG-RAN and 5G-NR: 5G Radio Access Network and Radio Interference

2020

GONTRAND Christophe
Digital Communication Techniques

PUJOLLE Guy
Software Networks: Virtualization, SDN, 5G and Security (2nd edition revised and updated)
(Advanced Network Set – Volume 1)

2019

LAUNEY Frédéric, PEREZ André
LTE Advanced Pro: Towards the 5G Mobile Network
Harmonic Concept and Applications

TOUNSI Wiem
Cyber-Vigilance and Digital Trust: Cyber Security in the Era of Cloud Computing and IoT

2018

ANDÍA Gianfranco, DURO Yvan, TEDJINI Smail
Non-linearities in Passive RFID Systems: Third Harmonic Concept and Applications

BOUILLARD Anne, BOYER Marc, LE CORRONC Euriell
Deterministic Network Calculus: From Theory to Practical Implementation

LAUNAY Frédéric, PEREZ André
LTE Advanced Pro: Towards the 5G Mobile Network

PEREZ André
Wi-Fi Integration to the 4G Mobile Network

2017

BENSLAMA Malek, BENSLAMA Achour, ARIS Skander
Quantum Communications in New Telecommunications Systems

HILT Benoit, BERBINEAU Marion, VINEL Alexey, PIROVANO Alain
Networking Simulation for Intelligent Transportation Systems: High Mobile Wireless Nodes

LESAS Anne-Marie, MIRANDA Serge
The Art and Science of NFC Programming
(Intellectual Technologies Set – Volume 3)

2016

AL AGHA Khaldoun, PUJOLLE Guy, ALI-YAHIYA Tara
Mobile and Wireless Networks
(Advanced Network Set – Volume 2)

BATTU Daniel
Communication Networks Economy

BENSLAMA Malek, BATATIA Hadj, MESSAI Abderraouf
Transitions from Digital Communications to Quantum Communications: Concepts and Prospects

CHIASSERINI Carla Fabiana, GRIBAUDO Marco, MANINI Daniele
Analytical Modeling of Wireless Communication Systems (Stochastic Models in Computer Science and Telecommunication Networks Set – Volume 1)

EL FALLAH SEGHROUCHNI Amal, ISHIKAWA Fuyuki, HÉRAULT Laurent, TOKUDA Hideyuki
Enablers for Smart Cities

PEREZ André
VoLTE and ViLTE

2015

BENSLAMA Malek, BATATIA Hadj, BOUCENNA Mohamed Lamine
Ad Hoc Networks Telecommunications and Game Theory

BENSLAMA Malek, KIAMOUCHE Wassila, BATATIA Hadj
Connections Management Strategies in Satellite Cellular Networks

BERTHOU Pascal, BAUDOIN Cédric, GAYRAUD Thierry, GINESTE Matthieu
Satellite and Terrestrial Hybrid Networks

CUADRA-SANCHEZ Antonio, ARACIL Javier
Traffic Anomaly Detection

LE RUYET Didier, PISCHELLA Mylène
Digital Communications 1: Source and Channel Coding

PEREZ André
LTE and LTE Advanced: 4G Network Radio Interface

PISCHELLA Mylène, LE RUYET Didier
Digital Communications 2: Digital Modulations

2014

ANJUM Bushra, PERROS Harry
Bandwidth Allocation for Video under Quality of Service Constraints

BATTU Daniel
New Telecom Networks: Enterprises and Security

BEN MAHMOUD Mohamed Slim, GUERBER Christophe, LARRIEU Nicolas, PIROVANO Alain, RADZIK José
Aeronautical Air–Ground Data Link Communications

BITAM Salim, MELLOUK Abdelhamid
Bio-inspired Routing Protocols for Vehicular Ad-Hoc Networks

CAMPISTA Miguel Elias Mitre, RUBINSTEIN Marcelo Gonçalves
Advanced Routing Protocols for Wireless Networks

CHETTO Maryline
Real-time Systems Scheduling 1: Fundamentals
Real-time Systems Scheduling 2: Focuses

EXPOSITO Ernesto, DIOP Codé
Smart SOA Platforms in Cloud Computing Architectures

LARRIEU Nicolas, VARET Antoine
Rapid Prototyping of Software for Avionics Systems: Model-oriented Approaches for Complex Systems Certification

MELLOUK Abdelhamid, CUADRA-SANCHEZ Antonio
Quality of Experience Engineering for Customer Added Value Services

OTEAFY Sharief M.A., HASSANEIN Hossam S.
Dynamic Wireless Sensor Networks

PEREZ André
Network Security

PERRET Etienne
Radio Frequency Identification and Sensors: From RFID to Chipless RFID

REMY Jean-Gabriel, LETAMENDIA Charlotte
LTE Standards
LTE Services

RIGO Michel
Formal Languages, Automata and Numeration Systems 1: Introduction to Combinatorics on Words
Formal Languages, Automata and Numeration Systems 2: Applications to Recognizability and Decidability

TANWIR Savera, PERROS Harry
VBR Video Traffic Models

VAN METER Rodney
Quantum Networking

XIONG Kaiqi
Resource Optimization and Security for Cloud Services

2013

ASSING Dominique, CALÉ Stéphane
Mobile Access Safety: Beyond BYOD

BEN MAHMOUD Mohamed Slim, LARRIEU Nicolas, PIROVANO Alain
Risk Propagation Assessment for Network Security: Application to Airport Communication Network Design

BERTIN Emmanuel, CRESPI Noël
Architecture and Governance for Communication Services

BEYLOT André-Luc, LABIOD Houda
Vehicular Networks: Models and Algorithms

BRITO Gabriel M., VELLOSO Pedro Braconnot, MORAES Igor M.
Information-Centric Networks: A New Paradigm for the Internet

DEUFF Dominique, COSQUER Mathilde
User-Centered Agile Method

DUARTE Otto Carlos, PUJOLLE Guy
Virtual Networks: Pluralistic Approach for the Next Generation of Internet

FOWLER Scott A., MELLOUK Abdelhamid, YAMADA Naomi
LTE-Advanced DRX Mechanism for Power Saving

FERRANT Jean-Loup *et al.*
Synchronous Ethernet and IEEE 1588 in Telecoms: Next Generation Synchronization Networks

MELLOUK Abdelhamid, HOCEINI Said, TRAN Hai Anh
Quality-of-Experience for Multimedia: Application to Content Delivery Network Architecture

NAIT-SIDI-MOH Ahmed, BAKHOUYA Mohamed, GABER Jaafar, WACK Maxime
Geopositioning and Mobility

PEREZ André
Voice over LTE: EPS and IMS Networks

TERRE Michel, PISCHELLA Mylène, VIVIER Emmanuelle
Wireless Telecommunication Systems

2012

AL AGHA Khaldoun
Network Coding

BOUCHET Olivier
Wireless Optical Communications

DECREUSEFOND Laurent, MOYAL Pascal
Stochastic Modeling and Analysis of Telecoms Networks

DUFOUR Jean-Yves
Intelligent Video Surveillance Systems

EXPOSITO Ernesto
Advanced Transport Protocols: Designing the Next Generation

FOULADGAR Javad
Electrothermics

JUMIRA Oswald, ZEADALLY Sherali
Energy Efficiency in Wireless Networks

KRIEF Francine
Green Networking

PEREZ André
Mobile Networks Architecture

POLER Raúl, CARNEIRO Luis Maia, JASINSKI Thomas, ZOLGHADRI Marc, PEDRAZZOLI Paolo
Intelligent Non-hierarchical Manufacturing Networks

2011

BONALD Thomas, FEUILLET Mathieu
Network Performance Analysis

CARBOU Romain, DIAZ Michel, EXPOSITO Ernesto, ROMAN Rodrigo
Digital Home Networking

CHABANNE Hervé, URIEN Pascal, SUSINI Jean-Ferdinand
RFID and the Internet of Things

GARDUNO BARRERA David, DIAZ Michel
Communicating Systems with UML 2: Modeling and Analysis of Network Protocols

LAHEURTE Jean-Marc
Compact Antennas for Wireless Communications and Terminals: Theory and Design

PALICOT Jacques
Radio Engineering: From Software Radio to Cognitive Radio

PEREZ André
IP, Ethernet and MPLS Networks: Resource and Fault Management

RÉMY Jean-Gabriel, LETAMENDIA Charlotte
Home Area Networks and IPTV

TOUTAIN Laurent, MINABURO Ana
Local Networks and the Internet: From Protocols to Interconnection

2010

BEGAUD Xavier
Ultra Wide Band Antennas

CHAOUCHI Hakima
The Internet of Things

FRIKHA Mounir
Ad Hoc Networks: Routing, QoS and Optimization

KRIEF Francine
Communicating Embedded Systems / Network Applications

2009

CHAOUCHI Hakima, MAKNAVICIUS Maryline
Wireless and Mobile Network Security

VIVIER Emmanuelle
Radio Resources Management in WiMAX

2008

CHADUC Jean-Marc, POGOREL Gérard
The Radio Spectrum

GAÏTI Dominique
Autonomic Networks

LABIOD Houda
Wireless Ad Hoc and Sensor Networks

LECOY Pierre
Fiber-optic Communications

MELLOUK Abdelhamid
End-to-End Quality of Service Engineering in Next Generation Heterogeneous Networks

PAGANI Pascal *et al.*
Ultra-wideband Radio Propagation Channel

2007

BENSLIMANE Abderrahim
Multimedia Multicast on the Internet

PUJOLLE Guy
Management, Control and Evolution of IP Networks

SANCHEZ Javier, THIOUNE Mamadou
UMTS

VIVIER Guillaume
Reconfigurable Mobile Radio System

Printed and bound by CPI Group (UK) Ltd, Croydon, CR0 4YY

13/09/2026

14961013-0001